33
PHOTOVOLTAIC
PROJECTS

Other TAB books by the author:

33 PHOTOVOLTAIC PROJECTS

BY HOMER L. DAVIDSON

TAB BOOKS Inc.
BLUE RIDGE SUMMIT, PA. 17214

FIRST EDITION

FIRST PRINTING

Copyright © 1982 by TAB BOOKS Inc.

Printed in the United States of America

Library of Congress Cataloging in Publication Data

Davidson, Homer L.
    33 photovoltaic projects.

    Includes index.
    1. Solar cells—Amateurs' manuals.
I. Title. II. Title: Thirty-three
photovoltaic projects.
TK9918.D38 1982   621.31′244   82-5926
ISBN 0-8306-2467-8   AACR2
ISBN 0-8306-1467-2 (pbk.)

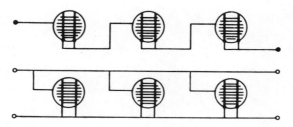

Contents

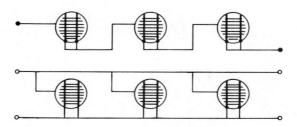

Introduction

Solar cells are exciting and fun to work with. They come in all shapes and sizes. The solar cell converts sunlight into electrical energy. A simple electromotive force is generated between the front and back surfaces of the cell when light strikes the sensitive front surface. Although the solar cell may sometimes appear weak and brittle, with a little extra care and treatment solar experiments offer a new electronic building experience.

Today, solar power is working in remote mountain communications, in navigational aids, railroad date transmissions, lighting commercial signs, and pumping water to livestock. Since solar heating and electricity has come into its own during the past few years, the solar cell now takes the place of some batteries and ac power supplies. Although the photovoltaic cell has come a long way, it is now opening up a new field of solar experimentation and solar powered projects.

This book will show you how in detail to build thirty-three solar electronic projects. Most of them are very basic and simple to assemble. Troubleshooting tips are given at the end of each project. Just pick out the solar project you want to tackle and have a go at it! Solar experiments are not only interesting and practical but also provide an introduction into the world of solar electronics.

Many manufacturers and individuals have contributed circuits and experimental data for use in this book. Many thanks for their kindness and consideration for without them this book may never have gotten off the ground. And to my seven children (whose faces and hands have appeared in many electronic experiments and service articles over the past forty years) I dedicate this book—to Larry, Colleen, Bruce, Julie, Cathy, Don, and Nancy—who have made my life complete.

Chapter 1

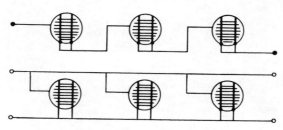

The Solar Cell

Solar cells are used to provide power to communication systems, water pumping, navigational aids, railroad warning lights, commercial lighting, and home electric systems. Originally, solar cells were expensive devices, used only in the space program. Since the beginning of the space age, the solar cell has become more practical and a whole lot cheaper. Today, many people are turning towards alternate forms of energy because of expensive oil prices. Many homes have some form of solar power or heating. Now, the photovoltaic cell has opened up a whole new world of solar electronic experiments for the average homeowner, beginning experimenter and technician.

The solar cell may be made up of germanium, selenium, or silicon. Most cells designed for electronic experiments are either selenium or silicon. The early solar cells used for powering radios and other small devices were selenium cells. Now, most all solar cells for modern electronic applications are silicon. The silicon cells are more dependable, cheaper, and more reliable than other cells.

When light strikes the surface of the silicon cell, current begins to flow from the negative pole to the positive. The top side of the cell, with grid-like lines, is negative with the bottom soldered-side positive. The voltage developed by the cell is .45 volts or less. You may find some cells listed at .5 volts. The current capacity of the cell is always equal to the physical size of the cell. The larger the size of the cell, the greater the current capacity.

This is a large four inch solar cell of the 2 amp variety. The larger the cell area the greater the current capacity. These cells are found in larger solar panels (courtesy of Radio Shack).

Solar cells come in many sizes and shapes, such as round, half round, one-quarter, square, or even broken pieces. Even small chips or broken crescent pieces of silicon cells will produce .45 volts dc. You may find silicon chips powering watches and calculators. Usually, these cells are only 1 to 6 mA types. To power a small portable radio the solar cell should have a current rating from 20 to 40 mA. The solar cell must be larger in size to provide the 100 to 400 mA of current required in some electronic projects.

Most solar power panels consist of many half-round or round 2-, 3-, or 4-inch solar cells. These panels may have an output voltage from 1.5 to 16Vdc. A 1.5 Vdc panel may have only 6 solar cells, while a 12 volt will have 32 cells and a 14 volt panel will have 36 individual solar cells. These panels may provide a current capacity from 400 mA to 2.5 amps (.9 to 40 watts). Greater voltage and wattage may be obtained by placing the solar panel in a series-parallel combination.

HOW TO CONNECT THE CELLS

To acquire higher voltages, the solar cells must be wired in series. When an electronic project requires 6 volt dc operation you

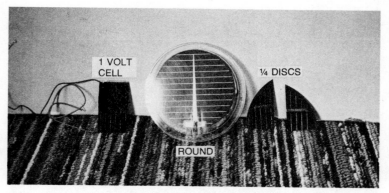

Solar cells come in many sizes and shapes. Here is shown a 1 volt two section type solar cell with a full round and two ¼ disc sections.

need to connect sixteen solar cells in series. Likewise, for a 12 volt operation you should connect thirty-two cells in series. You may find solar units listed with a great voltage output. Actually, a 1 volt cell consists of two cells with overlapped soldered connections. Three cells are soldered in series to form a 1.5 volt solar cell array.

Most electronic solar projects found in this book operate on 25 to 50 mA of current, although larger cells are used in solar panels. These cells may be connected in a series-parallel arrangement for greater voltage and current operations. For instance, if you had an electronics project that required 3 volts operation at 100 mills, you may use 50 mA solar cells. Simply connect two sets of seven cells in series. Then tie the two different strings of cells in parallel. Now

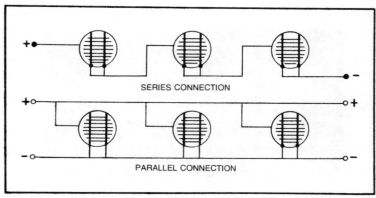

To acquire higher voltages the solar cell must be wired in series. Connect the cells in parallel to obtain higher current. Of course, the voltage does not increase in this particular parallel connection.

3

Two or more cells may be soldered together to provide greater voltage. Here three cells come in a 1.5 volt package with connecting leads.

you have a 3 volt-100 mA solar power source. The same voltage and current requirements may be met by connecting seven 100 mA cells in series.

The solar cell may come with color coded leads. Usually, the red lead (+) connects to the bottom and the black lead (−) to the top of the cell. You may find some cells with tab (flat metal type) leads. Extreme care must be observed when soldering wire connections to the sides of the solar cell. Too much pressure applied with the

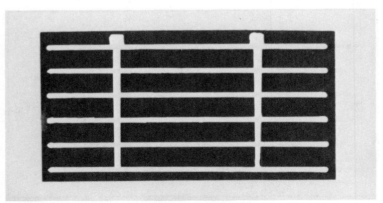

Most solar projects found in this book operate from 25 to 30 mA of current. These smaller cells may be crescent, quarter, or rectangular solar cells (courtesy of Radio Shack).

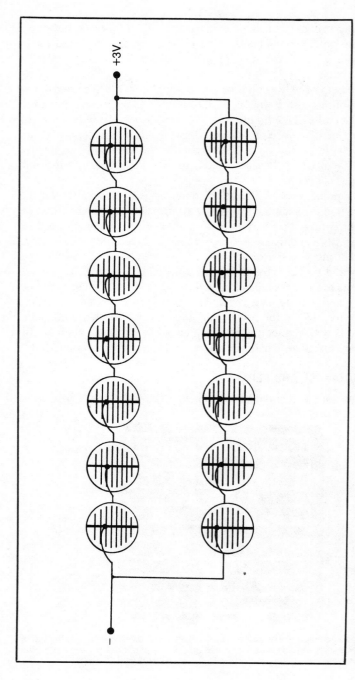

Fourteen 50 mA cells are connected in a series parallel arrangement for a 3 volt 100 mA solar power source. Solar cells may be paralleled to provide greater current.

soldering iron may crack or damage the cell. Too much heat may melt the soldered connection onto the glass base area. In many cases you will have to connect leads to cells that have no tab or wire connections.

To help keep the cell level, do not apply too much solder to the bottom of the cell. If you apply too much solder at the connecting point the cell will not lie in a level position. Keep all solder joints to a minimum. Use very thin connecting wire on all small solar cells. The solder must flow upon the cell connection without damaging the cell surface. Use a low power (25 to 50 watt) soldering iron for soldering connections.

Wrapping wire is ideal for connecting these small cells together. Radio Shack has 30 gauge Kynor wrapping wire in three different colors. You may take a section of ac rubber zip cord or speaker cable and make some very inexpensive connecting wires. Cut off a piece and remove the rubber insulation. Separate the strands and tin the wire ends for solar cell connection. The small wire does the job for connecting cells under 100 mills of current. For larger current cells, use regular hookup wire or metal tabs. Flat nickel wire is ideal for connecting larger cells. You will find some solar cells with connecting leads which are too long and must be cut off. Keep all these small pieces for future solar cell connections.

CELL BREAKAGE AND REPAIR

The solar cell should always be handled with care. These cells

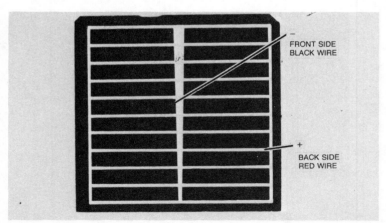

FRONT SIDE
BLACK WIRE

BACK SIDE
RED WIRE

Usually, a red lead (+) connects to the bottom and a black lead (−) to the top of the solar cell. When no leads are found, solder a lead to the top center bar area and the soldered back side of the cell (courtesy of Radio Shack).

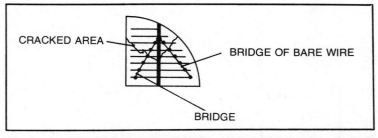

CRACKED AREA

BRIDGE OF BARE WIRE

BRIDGE

A broken or cracked cell may be repaired by soldering bare wire across the back side (soldered area) and the front of the cell. Bridge the broken areas with fine bare wire.

have a glass base and are easily broken. You will find the bottom side covered with a layer of solder. Between the bottom and top side is a layer of very thin glass material. The top side is the silicon layer. When mounting the solar cell make sure it is mounted level. Do not press down too hard on the cell. Keep extra cells between two pieces of cardboard or foam material. They cannot be thrown around or they will break.

Broken cells may be repaired or used for smaller current projects. A broken round cell may be repaired with solder applied to the back side. Solder the cell in several places. You can also bridge the broken area with fine pieces of wire. Repair the top side if it is cracked. The top side may be repaired by bridging small pieces of wire across the cracked area. The metal bar may be bridged with a bare piece of hookup wire. Go across the broken areas with a fine piece of wire. Form small beads upon thin grid lines of the top side. When a cell breaks into several pieces, it's best to use the separate cells for lower current carrying projects.

CUTTING SOLAR CELLS

One-quarter or crescent cells may be broken for use as lower current solar cells. For instance, a crescent cell of 100 mills may be broken in half producing two 50 mA solar cells. Do not try to break cell of the half-round or completely round type. You will end up with too many small pieces. Why would you want to break solar cells in two in the first place? The answer is availability and lower priced cells.

Let's take the case of one crescent cell (130 mills) at a cost of $1.50. Many of the solar projects in this book operate with less than 50 mills of current. By cutting the cell in two you now have two 65 mA cells at a cost of 75¢ each. You cannot buy them at a lower price.

7

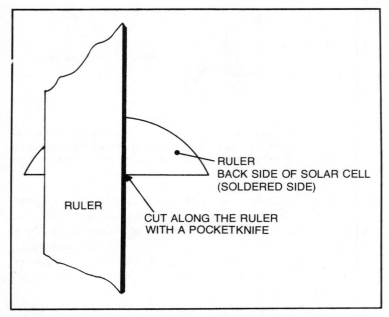

RULER
BACK SIDE OF SOLAR CELL
(SOLDERED SIDE)

RULER

CUT ALONG THE RULER
WITH A POCKETKNIFE

One-quarter or crescent cells may be broken for lower current projects. Use a ruler or soldering iron for this purpose.

Remember it takes at least two cells in series to make 1 volt. With this process you get two cells for the price of one.

Today, most solar cells are professionally cut with a laser beam. Since laser beams are not found in the ordinary workshop, you do the next best thing. The cells may be cut and broken in two with a soldering iron and ruler. In one project, six 225 mA ¼ inch cells were cut in two for a project calling for twelve 100 mA solar cells. These cells were broken in half with a pocket knife and ruler. Lay the cell face down on a board or heavy cardboard. Lay the ruler half way across the center of the cell. Lightly cut a path across the soldered area. Keep applying more pressure until a thin line is cut into the soldered area. The cell will break down this cut area.

Another method is to cut and break the cell with an electric soldering gun. Remove the solder U-tip. Make a flat type tip out of number 14 copper wire. Simply strip the insulation off of a piece of solid 14 or 12 gauge copper wire. Form the wire so a flat area will go over the center of the solar cell. Place the cell on a flat surface. Press the iron tip down onto the back side of the cell and turn the gun on. Apply pressure downward as the solder begins to melt.

Six 225 mA-¼ inch cells were cut for a project calling for twelve-100 mA solar cells. These ¼ inch sections came from a three inch cell.

Apply a pocket knife blade to the cutting area if the cell doesn't break in two. Usually, the cells are easily broken at this point. Keep the cells for future lower current projects when the cell breaks into several pieces. Don't throw any small cell pieces away. You may need them in a later solar experiment.

MOUNTING THE CELLS

The solar cell may be mounted onto any piece of material that will not short out the back side of the cell. When mounting cells onto a metal surface, place a layer of plastic or plastic screen behind the soldered cell area. Do not lay the bare cells down on the bare metal chassis. If you are going to use a metal chassis or mini-box to house

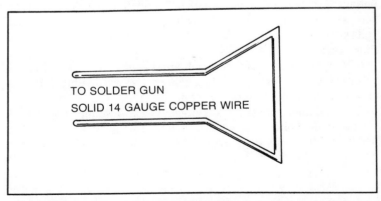

TO SOLDER GUN
SOLID 14 GAUGE COPPER WIRE

Solar cells may be broken with a homemade tip for the soldering gun. Make a flat type tip out of solid 14 gauge copper wire.

your favorite project, mount the cells on a piece of plastic or leave the cells in the plastic box and cement the plastic box to the metal.

Solar cells may be mounted on wood, masonite, or foam material. All of these materials provide adequate insulation for low voltage solar cells. Plastic foam is ideal to mount the cells on. The foam is easy to cut and provides a safe mounting area. Usually, the foam material may be found in packing boxes, picked up at a lumber store or hobby shop. Foam material is fairly inexpensive compared to other insulating materials.

The cells may be cemented to the insulating base with silicone rubber cement, epoxy, or regular cement. Silicone rubber cement does a nice job and provides a cushion for the solar cell. Apply enough rubber cement to evenly cover the back of the cell. This includes covering the entire area at least even with the cell connecting wire. Remember the bottom side of the cell connecting wire must be soldered before applying cement to the cell area. Black cement leaves a smudged and dirty appearance. The cells should be lined up in a few minutes since the rubber cement will set-up in ten to fifteen minutes, depending on the temperature.

COVERING THE CELLS

Before covering the cells (for added protection) use a vom and check the output voltage. It's very difficult to repair or replace a cell which is covered with rubber cement or epoxy. Besides having a poorly soldered connection, you may accidentally crack or break one or more cells. Sometimes these hairline cracks are difficult to see. A voltage output test will reveal if you have a poor voltage output. To locate the defective cell or connection, check the voltage across two different cells. The voltage across the two individual cells should be the same. Use the 1 or 3 volt dc range of the vom for these tests.

Remember the maximum voltage output of a solar cell is .45 volts under ideal sunlight. When using a 100 watt bulb for testing purposes, the output voltage may measure less than three-fourths of the maximum voltage. The voltage will be less than one-half of the maximum voltage output if the cells are strung out where only a portion of the cells are covered by the light bulb. For projects operating with low voltage and current, the reading lamp or 100 watt testing light bulb provides adequate lighting for the cells. Solar panels and heavy current experiments must be operated under sunlight.

Check the voltage output of the solar cells before sealing them. The cells may be checked by using either sunlight or a 100 watt bulb.

To protect the solar cells from breakage, dirt, dust, and weather some kind of protection should be added over the brittle cells. Usually, for outside protection, solar panels are covered with plastic or are molded right inside the plastic panel. The solar panel must be protected from all types of weather. When the cells are used indoors they need to be protected from possible breakage. Clear plastic material is used in most cases. You may cover the entire solar area with a plastic box or a piece of flat plastic. Look around the house for small plastic boxes. Today, just about everything comes in plastic boxes. Some solar cells come packed inside a plastic box. These are ideal to mount on the top or at the back of your electronic solar project. A piece of foam rubber is usually used for packing and protects the cells if the project is dropped or slides off the workbench.

Separate lids of plastic boxes may be used to cover up the cells. Of course, the plastic container must be large enough to cover up all of the cells. The plastic lid may be cemented to the project case. Use silicone rubber cement for this purpose. When cementing a plastic lid to a black plastic box use black silicone rubber cement. Otherwise, use clear or white rubber cement for a cleaner appearance.

Plastic picture frames provide excellent protection and mounting frames for solar voltage panels. These plastic frames come in 5 × 7 up to 16 × 20 inch size. They are made of heavy plastic and help

11

to prevent cell breakage. The solar cells may be mounted on a foam area. Just cut the foam material the same inside dimensions as the plastic picture frame. The plastic edge of the picture frame is strong enough for mounting voltage take-off jacks and dog-ears.

The solar cells may also be covered with small pieces of clear plastic. Some solar panels come with a lens-type plastic covering. Although, this lens-type plastic may not be available in your area, various sheets of plastic may be found at a lumber yard or discount store. Choose a clear sheet of plastic. Do not use colored plastic since it will reduce the amount of light. Plastic P4 lens material is ideal to provide additional light to the solar cell surface.

Liquid plastic resin and silicone rubber cement may be used over the entire solar cell area. Some commercial solar panels are embedded inside plastic material, while others have a special low iron, water-white grade of tempered glass. This special glass transmits a higher percentage of light energy than regular plate glass. Clear rubber silicone cement may be used over and around the solar cell surface for added protection. Solar panels used out-doors should be sealed with liquid plastic resin or silicone rubber cement.

CHOOSING THE CORRECT SOLAR CELL

The solar projects in this book list the correct solar cells to properly operate the experiment. When designing or building your own solar project, you must know the operating voltage and current of the particular device to be powered with solar cells. For instance, you cannot run a small dc motor with only a 10 mA solar cell. Likewise, a colored portable TV receiver cannot be operated from a 12 volt 1 amp solar panel. You must know the voltage and operating current of the project involved.

Let's say you have a small portable radio you wish to operate from solar power. The service literature may list the battery voltage and current. But, it may not. So, simply count the number of small batteries. If there are four penlight cells found in the battery compartment, since the penlight cells are 1.5 volts each, we come up with an operating voltage of 6 Vdc. If in doubt, measure the voltage with a vom.

To determine how much current the radio is pulling from the batteries we must measure the current with the radio in operation. Insert a small piece of cardboard between one battery located at the radio terminal connections. Clip the vom leads across the two battery contacts separated with the cardboard. Actually, you are

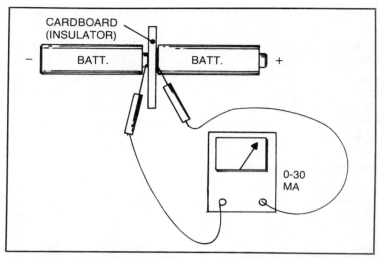

CARDBOARD
(INSULATOR)

− BATT. BATT. +

0-30
MA

Before attempting to solarize your favorite project or unit, take a current reading. Place the vom leads in series with the batteries. Set the vom to the mA dc range.

placing the vom leads in series with the batteries. You may accomplish the same thing by unsoldering one battery lead to the radio chassis. Set the vom to the 30 mA dc scale. Now, turn on the radio.

Notice if the meter hand goes backwards. Reverse the meter leads if the hand goes to the extreme left. As you turn up the volume of the tuned in station, notice how the meter moves up and down the scale. Increased volume moves the current meter hand farther up the scale. You will find most portable radios operate on 10 to 30 mA of current. You may use cells below 50 mA of current for most portable battery applications.

Let's assume the portable radio pulls 20 mA of current and operates from a 6 volt source. We can use 25 mA solar cells for this project. Fourteen (25mA) solar cells must be connected in series to acquire the operating voltage (6 Vdc). Always, add a couple of extra solar cells in the circuit for adequate sunlight operation. The current and operating voltage of any project or device must be known before solar power may be applied.

Lower current project applications may be operated from crescent or broken cell pieces. These broken cells may be purchased in a 10 piece lot for only a few dollars. Today, this means many solar projects may be powered for less than ten dollars. Many of the projects in this book are operated from broken pieces and crescent solar cells.

For solar projects between 100 and 500 mA, choose either half of one-quarter round discs. These cells are cut from 2 to 4 inch round cells. You do not want to use full round cells for lower current projects. The one-quarter or half round cells are much cheaper and take up a lot less space than the full round cells. In fact, you may purchase 4 or 6 one-quarter discs for the price of one large cell. Keep the project price down by choosing the right solar cell. Choose only the full round cells when current requirements are 1 amp or greater.

BUILDING THE SOLAR PANEL

Today the solar panel is mounted outside and powers many devices. The solar panel is ideal to charge up a storage battery. Several solar panels may be parallel for home electric power systems. Typical home electric power systems may consist of four solar panels, three or more 12 volt storage batteries, regulated control panel, power cables, and mounting hardware.

For lower current and multi-voltage panels used to power transistor radios, toys, motors, or any low voltage project, you may either build or purchase a commercial solar panel. These commercial 50 mA multi-voltage panels cost less than twenty dollars. Project 1 describes how to build a 6 volt - under 50 mA voltage cube source. A larger current solar panel (100mA) with multi-voltage taps may be constructed as described in Project 16.

Commercial 12 or 14 volt solar panels may cost from $150.00 to over $1,000.00. The cost of the panel depends upon the operating voltage and amperage of the solar panel. Several of these solar panels must be operated in parallel for a home-electric system. A 12 or 14 volt solar panel may be used for charging auto batteries. Project 31 shows you how to build a Delux Solar Battery Charger. A camper 12 volt panel is described in Project 33 while a commercial solar panel kit may be constructed for less than $300.00 as shown in Project 32.

Before building or purchasing a solar panel, make certain that the finished project will meet the correct requirements. Lower voltage and current operating panels may be constructed for less than commercial units. For higher voltage and current operation the cost of either building or purchasing a commercial solar panel may be about equal. One thing is for certain, the experience and gratification of constructing your own solar panel cannot be measured in dollars and cents.

This is a Solarex large cell, high density, solar panel using square cells. 36 large 85 mm cells provide a nominal 14 volts at 2.1 amps and provide 34 watts peak (courtesy of Solarex Corp.).

THE SOLAR CELL AND ELECTRONIC SYMBOLS

The solar cell symbol may be new to you. This symbol and the solar cell drawings used in this book may be referred to for making solar cell connections. Check for correct polarity before connecting the solar cell to the circuit. Generally, the top side of the cell is negative and the bottom side is positive. The voltage output of each cell is less than .5 volts dc.

When making integrated circuit (IC) component connections, only the IC terminal connections are found in the schematic diagrams. The internal functions of components found in the IC would only add confusion to building the projects in this book. All IC connections are from the *top side* of the IC. The best way to start wiring an IC socket is to start with terminal 1. This terminal corresponds with a dot mounted on top of the IC component. Do not plug the IC into the circuit until all terminal connections are completed.

Many of the solar projects have a pictorial diagram for easy wiring connections. For those without any electronic construction experience, the solar project may be connected by following the pictorial diagram. Always, check out the wiring connections and

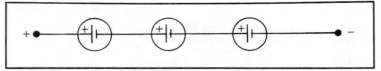

The solar cell symbol may be new to you. Actual solar cell drawings are used throughout this book.

components before firing up the project. It's very disheartening to complete a project and nothing works.

TROUBLESHOOTING A PROJECT

No matter how hard a person tries to follow the construction procedures, once in a while a project will be completed and may not function. You will find a troubleshooting procedure at the end of most of the solar projects. Here, various test points and tips are given to find the possible trouble spot.

The solar cells may be checked with the low voltage scale of the vom. Remember, the cells must be under a 100 watt bulb or sunlight to produce voltage. When lower voltage than normal is suspected from the solar cells, check for a broken wire, poorly soldered connections or a cracked cell. Check the voltage across two cells until you have located the defective area. A single cell should be tested on the low voltage dc scale of the vom. Under direct sunlight, a solar cell should produce .45 volts.

Continuity tests may be made with the low R×1 vom scale. The low ohm scale will locate a broken connection, open coils, broken low ohm resistors, and open cable wires. Set the vom to the R×1 scale and place the two test leads across the component or wires to be tested. A high or no resistance reading indicates an open component or connection.

A fixed silicon or zener diode may be checked for an open or leak with the vom. Set the vom to R×1 scale. Place the two resistance probes across the suspected diode, you will have a low reading (20 to 50 ohms) in one direction if the diode is normal. Reverse the test probes at the diode and a normal diode will not have a reading. Now turn the vom up to the R×10 scale with the probes attached across the diode and no reading indicates a good diode. A leaky diode will show a reading in both directions. When no reading is obtained in any direction, the diode is open.

You may check for a leaky or open transistor in the very same manner. Set the vom to R×1. Place the positive (red) lead to the base terminal. A low ohm reading will be noted when the black

16

probe touches the emmitter or collector terminals. No ohm reading indicates the transistor is open. Now, reverse the procedure. Place the black probe to the base terminal. A normal pnp transistor will not show any reading on the emitter or collector terminal. When a low reading is noticed between emitter and collector terminals, the transistor is leaky. Notice that these tests are made only on the type of pnp transistor used in the projects in this book.

You can check the condition of an electrolytic capacitor with the R×10 scale. The electrolytic capacitor must be out of the circuit for these tests. Hold one end of the capacitor with the probe tip and quickly touch the other capacitor lead. Notice how the meter hand goes up and back. Reverse the test leads. The capacitor should charge up and back if normal. You will notice that the larger the capacitance, the greater the swing of the meter hand. A no-charge up with a low resistance reading indicates a leaky capacitor. No charge-up of the meter hand indicates the capacitor has lost it's capacitance. The smaller the capacitor the less the distance that the meter hand will move.

Most problems found in a non-working project are poor solder joints or improper connections of various components in the circuit. Usually, these problems can be eliminated with a little care and patience. Proper voltage and resistance checks with the vom may help locate these troublesome areas. Good soldering techniques are important to follow when building the solar electronics projects.

TEST EQUIPMENT AND TOOLS

Practically every home or workshop has adequate tools for these solar projects. A low wattage soldering iron is a must around

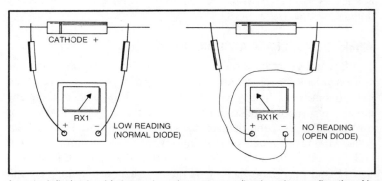

A normal diode should show a low ohmmeter reading in only one direction. No reading at all indicates an open diode. Any type of resistance reading in both directions indicates a leaky diode.

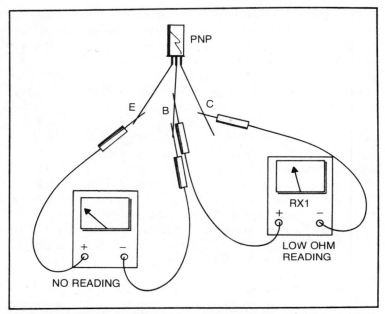

The pnp transistor used in the book may be checked with the RX1 scale of the vom. Place the positive (red) probe to the base terminal. A normal transistor will have a low ohm reading between either base and collector or base and emitter. Reverse the leads with the black probe at the base. A normal pnp transistor will have no reading even on the higher ohmmeter scales.

electronic components. A pair of long nose pliers with side cutters is also a very handy tool. Several small screwdrivers round out the tool requirements. The low priced vom is the only test instrument needed. A low priced vom can be purchased for ten to about thirty dollars.

WHERE TO PURCHASE A SOLAR CELL

Several years back only a few selenium and silicon solar cells were available for electronics projects. Many of these were of the low current types. Today, you may purchase solar cells in about any electronics supply house. These solar cells may be cracked or broken cells, crescent, quarter discs, half round, or full round solar cells. Mail order houses supplying low priced broken or crescent cells are given in the Suppliers section at the end of this book.

Higher current cells may be quarter disc, half round, and full round solar cells. The cell sections are taken from full solar cells. Before purchasing any solar cells, check out the various prices and

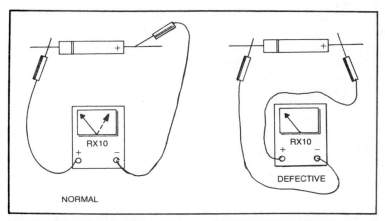

NORMAL

RX10

DEFECTIVE

RX10

Use the ohmmeter and "kick test" for small electrolytic capacitors. A normal electrolytic capacitor will cause the meter hand to fly up and back. A no "kick" reading or a low resistance reading between the capacitor terminals indicates that the capacitor is leaky.

requirements. The Suppliers section lists places to obtain these type of cells. Many of these same firms have surplus manufacturer's cutoffs, and overrun cells for less money.

The same electronic firms who carry solar cells may have electronic components to build the various solar projects. You will find a parts list at the end of each project with part numbers and names of electronics firms which sell the specific components. This

Only a handful of tools are needed to build the solar electronics projects in this book. A low wattage soldering iron is a must item when soldering around small solar cells and components. A low priced vom checks voltage, current, and continuity in the circuits.

does not mean you cannot pick up the part locally or use it out of the parts bin. Don't forget to check the back pages of electronic magazines for solar cells and electronics components. They just may have the exact part you have been looking for. Now, let's pick out a solar electronics project from the following pages and begin to have some fun.

Chapter 2

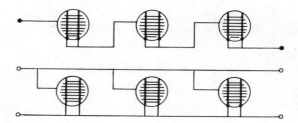

Thirteen Projects for Under $25

This little solar voltage cube was constructed using crescent solar cells. Actually, these cells are outside sections from round solar cells. The voltage from the cells is .45 V with a 15 to 36 mA current rating. An operating voltage of 5.5 volts is accomplished with thirteen crescent cells wired in series (Fig. 1-1).

The solar voltage cube is an electronics experimenter's delight. Here you may operate a small radio, motor, solar experimenter's kit, or pocket calculator from this solar voltage source. If you desire a greater voltage source, just add a few more solar cells. Any combination of solar cells may be used to acquire the correct current and voltage supply. The cost of this project is under twenty dollars.

Connecting the Cells

The thirteen solar cells must be wired in series to acquire the needed 5.5 volts. Use a strand of copper wire to connect the cells. Scrape off the junction at the largest vein running through the center of the crescent cell (Fig. 1-2). Tin the bottom side of the cell with solder. Now, apply solder with a small soldering iron to make a bead of solder on the top side of the cell. Solder a small strand of wire to

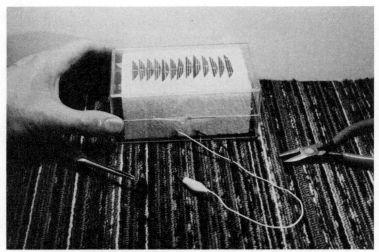

Fig. 1-1. Here is the finished solar cube. Thirteen crescent solar cells are wired in series to produce a 5.5 Vdc voltage source.

the drop of solder on the top side surface. These strands of copper wire may be taken from a section of ac cord.

Turn the cell over. Now, lay another cell ⅛ inch from the first solar cell. Solder the piece of wire to the bottom side (positive). Cut off the excess wire with a pair of side cutters. The bottom side of the cell solders up nicely since the whole area readily takes solder. Do not apply too much heat from the soldering iron. Now, turn the cell over and start once again.

You may want to scrape and tin each top side (negative) connection before connecting the next cell. At least it's best to tin each top side area before soldering in place. Only a drop or bead of

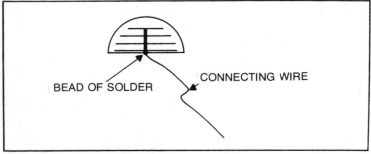

Fig. 1-2. Scrape off the top junction with a pocketknife. Place a bead or drop of solder on this junction to connect the cell tie wire.

solder is needed for the top side connection. The cells are easily turned over by simply taking hold of the two connection wires (on each end) and flipping the cells over. Before you know it all thirteen cells are wired in series. Try to keep the small connecting wires in the center of the crescent cells.

The copper connecting wires should be tinned with solder. This is easily accomplished by placing a little excess solder on the iron tip and drawing the small wire through the liquid solder. Apply solder at the same time. All sides of the small copper wire may be tinned at the same time. If the wire will not tin properly, place solder paste between your thumb and finger. Now pull the small copper wire through. The copper wire may usually be tinned without any problems (Fig. 1-3).

Mounting the Cells

After all cells are wired in place, check the voltage under sunlight or a strong work light. Under strong sunlight, you should have over 5.5 volts dc. If not, a poor soldering connection or defective solar cell must be located. Measure the voltage across two separate cells until you have located the poor connection.

The space needed for the thirteen crescent cells should not take over four inches. Select a piece of foam material to mount the cells on (Fig. 1-4). To protect the cells from dust and damage, select a plastic box to place the foam cube in. This prevents cell breakage and provides a solid movable testing case. Cut the foam cube to fit nicely inside the plastic box. Actually, any plastic box will do as long as a cover is provided to enclose the solar cells.

Center the foam cube and draw a line down the center. Place a bead of silicone cement on this line. If too much cement is used, it

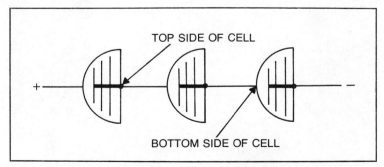

Fig. 1-3. Connect and solder the small wire to the top side of one cell and to the bottom side of the next cell. Space the cells ⅛ inch from each other.

will push up around and on top of the cells. Use a clear rubber silicone cement for holding the cells in place. Only a thin line of cement is needed. Lay the center of the thirteen cells on this line and embed the crescent cells into the cement. Press down so all cells are level. Scrape off any excess cement from the top of the cells. Let the cement set up over night.

Connecting the Clips

Before the solar cube is placed inside the plastic container, solder two flexible leads to the connecting wires. Use a red wire for the positive terminal (bottom of cell) and a black wire for the top side of the cell. Simply push a hole up from the bottom side of the foam cube on each side of the cells with an ice pick or awl.

Feed the flexible leads through the foam to connect to the cell leads. Solder the leads to the two cell wires. Now, pull the excess connecting wire down through the holes in the foam material. Taping the soldered connection is not necessary. The bare lead connection should be down inside the foam area.

With the tip of the soldering iron, push two small holes in the front of the plastic case for the two flexible leads. Feed the two wires through these holes and fit the foam cube inside the plastic box. Keep pulling on the excess lead wire until all slack is pulled through. Now, solder two alligator clips to the flexible leads (Fig. 1-5). Match up a red covered clip to the red lead and a black clip to

Fig. 1-4. Here is a top-side view of the crescent solar cells cemented to a piece of foam. Cement the cells in place with clear silicone rubber cement.

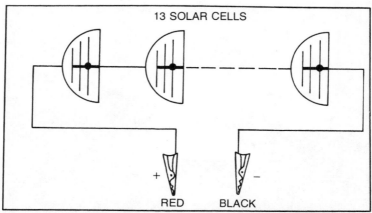

Fig. 1-5. All cells should be connected in series. For easy connections, solder two alligator clips to the flexible terminal wires.

the black wire or negative lead. If the crescent solar cube is to power or take the place of a certain battery, solder the appropriate plug instead of the alligator clips.

The amount of current and voltage from the solar cube is determined by the size of the solar cells. Small radios, calculators and some electronic experiments may be powered by cells producing only 36 or 40 mA of current. Although some small motors may operate with this amount of current, many of these motors may only operate with cells which produce over 125 mA of current. Select the appropriate cells for whatever operating voltages and current ratings are needed for your project.

☐ Construction time—2 to 3 hours.
☐ Cost of Solar Cube—$10 to $20.
☐ Available Solar Cells—Crescent solar kit, number P-42749
                                    Edmund Scientific Co.
                —Crescent solar cells - 225 amps
                   Solar Amp.

PROJECT 2
A SIMPLE SOLAR RADIO

Here is a simple broadcast radio you can build that will operate with solar cells under a reading lamp or in sunlight. You can listen to your favorite music or a football game while reading and will not bother anyone in the house. This little radio was constructed inside of a plastic Legg Hosiery container (Fig. 2-1).

Fig. 2-1. This is a photo of the simple solar radio. All of the components fit inside a Legg hosiery container.

How It Works

Actually, the radio circuit is basically a crystal radio with a two stage transistor amplifier (Fig. 2-2). The small flexible throw around antenna wire picks up the broadcast signal. L1 and C1 tune in your favorite local station. L1 picks up the tuned signal and is detected with the fixed crystal detector D1. The audio signal is coupled directly to the base terminal of Q1.

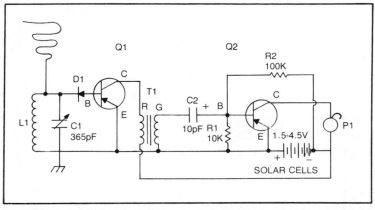

Fig. 2-2. The radio circuit consists of a crystal radio front end and two stages of transistor amplification. Several local broadcast band radio stations may be tuned in with this simple solar radio.

Q1 amplifies the weak audio signal and is coupled to Q2 with transformer T1. Q2 amplifies the audio signal to drive a small earphone. R1 and R2 are base bias resistors. The two transistors are powered by three or more solar cells. Greater volume may be obtained by adding more solar cells. If the volume is too great on local stations, simply turn the solar cells away from the light.

Component Availability

Practically all the parts should be available at your local electronics store to build this solar radio. You may have most of them in your electronics parts box. C1 and L1 may be a little more difficult to obtain locally, but can be ordered from the firms listed in the parts lists (Table 2-1). L1 may be a ferrite adjustable antenna coil or a fixed bar or rod type coil. As long as the coil will match with a 365 pF variable capacitor and will fit inside the plastic Legg container, use it.

If L1 is not handy you can wind your own (Fig. 2-3). Select a ⅜ inch diameter ferrite rod not over 3¼ inches long. Pick up some number 24 or 26 enameled wire to wind the coils. Sometimes this wire can be salvaged from old transformers, relays or other wound coils. Either enameled or cotton covered wire may be used to wind L1.

Start at one end of the ferrite rod and place a small strip of scotch or masking tape. Leave four inches of wire at the beginning for hookup to the variable capacitor. Tape down the end and wind the wire over the strip of tape. This will hold the winding in place as you begin to wind L1. Now close wind 72 turns on the coil form. When finished, secure the last few turns with tape. Always leave three or four inches of wire at each end to connect to the various components.

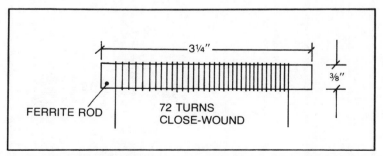

Fig. 2-3. If L1 is not handy, wind your own. Check the parts list (Table 2-1) for required data and components.

Table 2-1. Parts List for Project 2.

L1 — Ferrite antenna coil to match 365 pF variable capacitor.
Winding your own coil:
    1 — MS081 ferrite rod — ETCO Electronic Corp.
    7ft. #26 enameled wire — 72 turns over ⅜ inch ferrite rod.
C1 — 365 pF variable capacitors
    Part number — AL-233 GC Electronics
    Part number — 185VA — ETCO Electronic Corp.
C2 — 10 μF 10V electrolytic capacitor
R1 — 10K ½-watt carbon resistor
R2 — 100K ½-watt carbon resistor
D1 — 1N34 diode
T1 — Driver transformer
    Part number — DL-728 500 - 1,000 Ω GC Electronics
    Part number — 13A833.9 — Gravois Merchandiser Inc.
    Part number — 002XF — ETCO Electronic Corp. or equivalent.
Q1 & Q2 — 2SA52 or RCA SK-3004, GE-2, Radio Shack 276-2007 or
    276-1604 pnp Germanium
    24-1625 — GC Electronics transistors or equivalent.
P1 — 1000-2000 ohm impedance
    Part number J4 — 825 GC Electronics
    Part number 272VA —ETCO Electronics, Inc.
Solar cells — 1.5 to 4.5 volts — 36 to 50 mA
    2 — number —JA — 803 GC Electronics
    6 — number —S122 — Solar Amp
    1 kit — Solar Crescent cells number P-42 749 Edmund Scientific Co.
Misc. Leggs plastic hosiery container, 4 terminal wire strips, piece of foam or equivalent wood piece (for base), 10 ft. flexible wire for antenna, solder.
Construction time — 6 to 10 hours (or less).
Cost — Under $25.00 with parts from scrap box.

L1 may be a commercially built coil or a hand-wound one. The coil may be a flat or round type. It doesn't make any difference as long as it will physically fit inside the Legg container and matches C1. L1 and C1 may be salvaged from a small transistor radio as a matched pair. The variable tuning capacitor (C1) can be any size or shape. Of course, the smaller the better to fit inside the top section of the Legg container.

The earphone should have an impedance of 1,000 or 2,000 ohms. Do not use an 8 ohm earphone as found with most transistor radios. You can pick up the required earphone at most electronic parts stores. A crystal earphone may be used if you already have one, but be sure and place a blocking capacitor in series with one earphone lead.

Q1 and Q2 are low drain pnp transistors. Practically any germanium low signal transistor may be used here. Germanium transistors seem to operate very efficiently with low drain solar cells. With this circuit local broadcast stations can be heard with a 1.5 Vdc operating voltage. RCA, GE, and Radio Shack transistors work well in this type of circuitry.

The interstage transformer (T1) may have a primary impedance of 1K to 500 ohms. These small audio driver transformers come in many sizes, the smaller the better. Here one of the mounting brackets was soldered to the ground terminal lug for mounting. Of course, the transformer can be wired into the circuit, then mounted with a dab of silicone cement to one side of the plastic bottom section.

Mounting the Larger Components

The variable capacitor (C1) is mounted in the very top area of the Legg container. Drill or use a pencil iron and make a hole large enough for the shaft of the small capacitor. If the mounting hub of the capacitor is not long enough to apply a lock nut on the top side of the case, cement the capacitor into position with clear silicone cement. Place the knob on so that the shaft can rotate freely.

Line up the capacitor so that the middle of rotation will be at the top of the container. Of course, the top Legg shell may be rotated after the capacitor is mounted. Now, place cement on the two opposite corners. Rotate the capacitors once again to see if the capacitor will turn 180°. Let the cement set up overnight.

Check the antenna coil for correct length. If the coil is over 3¼ inches, cut off the excess ferrite core to mount in the bottom section. These ferrite round cores may be cut with a hacksaw and then broken off with a pair of pliers. A pair of large electronic pliers can snap or break off the ends without using the hacksaw blade for cutting the line. Place the side cutting edges of the pliers where the core is to be cut. Usually, these round ferrite cores will break at this point. If the coil is too long, the mounting end may be packed down through a mounting hole at the back side of the plastic container. Cut a slot or round hole just big enough for one end of the core to go through. Stand the coil up straight and place silicone rubber cement around this area to hold it in place.

The transistor and other small components are mounted on a four terminal wire strip. Cement the metal lug to the opposite side of the coil in the bottom plastic area. This keeps the small parts together and at the bottom section of the radio. After the three major components are cemented in position, let the cement set up overnight. All other components may be mounted as they are wired in the circuit (Fig. 2-4).

Wiring the Radio

Most components can be mounted as they are wired into the circuit. Connect the small transistor to the terminal strip and solder in the various resistors and capacitors. Use the center terminal or

Fig. 2-4. This is a photo showing all of the small components mounted and wired into the bottom of the egg section. Check off each component in the schematic as they are wired into the circuit.

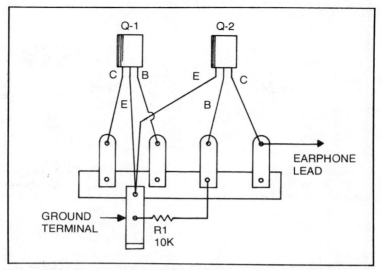

Fig. 2-5. Mount all small components on the four terminal strip. Use the long nose pliers as a heat sink when soldering the diode and transistor leads.

metal lug of the terminal strip as the common ground connection (Fig. 2-5). Keep the small components bunched down towards the bottom of the plastic container.

You may want to mount most of the small components on the terminal strip and then glue it into the bottom of the plastic section. This may save a lot of time and prevent the soldering iron from touching the small plastic case. Place all small parts into the bottom holes of the metal lugs after all other parts are soldered. Now, wire in the small crystal diode (1N34). Use long nose pliers as a heat sink while soldering the diode and transistor leads.

Drill or poke a hole in the bottom back side of the plastic container with the soldering iron for the earphone cord and antenna wire. Either tie a knot in the cord or place silicone rubber cement over it so the cord cannot be pulled out of the radio. After all parts are soldered into the circuit, wire the variable tuning capacitor and solar cells. Connect the long coil wires (L1) to the tuning capacitor after the solar cells are cemented to the top egg section (Fig. 2-6).

Powering the Radio

With earphone operation only a few solar cells are needed. The operating voltages may be anywhere from 1 to 4.5 volts. This can easily be obtained with commercially built cells or simply construct

31

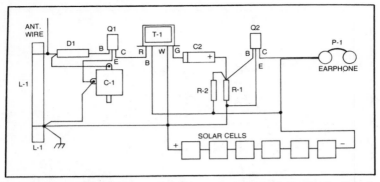

Fig. 2-6. Follow the pictorial wiring diagram for easy hookup. Check off each component as it is wired into the circuit.

your own (Fig. 2-7). To prevent breakage of the cells, they should be mounted on a piece of stiff plastic. In fact, the solar cells may be mounted upon or inside the small plastic container they are shipped in. For larger solar cells, cement them directly to the foam area.

To build your own, select four to six solar cells with a current rating of 36 to 60 mA. When wired in series, the voltage should be somewhere between 1.5 to 4.5 volts. If you desire more volume or if the local radio station is twenty-five or more miles away, you should add a few more cells in series for a greater voltage source. The higher the voltage, the more cells that are needed to be connected in series. Here, six small solar cells are connected in series (Fig. 2-8).

The bottom shell of the egg container may be cemented to the foam base before any parts are mounted in it (Fig. 2-9). A piece of wood may be used as the base instead of foam. Pick up a small plastic ring at one of the hobby stores. Place this under the point of the egg and cement all three pieces together with silicone rubber cement. Tilt the egg at an angle and straighten up with the bottom base piece. Allow the cement to set up overnight.

Checking the Radio

After all components have been wired, replace the top egg section. Now, place the unit in the sun or underneath a reading lamp. Rotate the tuning knob to your favorite local station. You should be able to tune in several radio stations. If nothing is heard, remove the top plastic section.

Measure the voltage with the vom at the solar cell connections. You should have from 1.5 to 4.5 volts depending upon how many

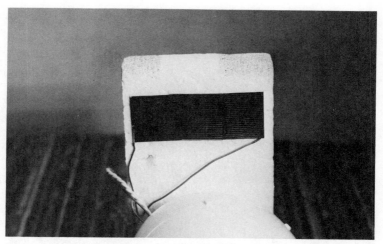

Fig. 2-7. Mount the solar cells on the top side of the base section. Drill or push a small hole into the plastic area for the solar cell wires.

cells are in series. If voltage is low, check the polarity of the solar cells. You may have them wired backwards. The top section of the solar cell (black) is negative and should go to the earphone junctions. The positive lead (red) should go to common chassis ground.

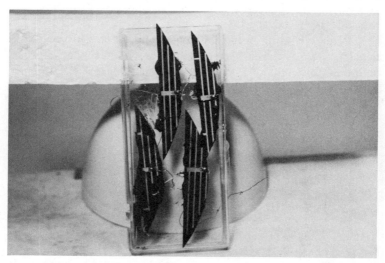

Fig. 2-8. Here are four small crescent solar cells mounted on the lid of a plastic box. These cells are connected in series.

Fig. 2-9. The bottom shell can be mounted on the foam base before any components are placed inside. Cement the shell to the base with silicone rubber cement.

If these connections are correct, go over each component and see if they are wired in correctly. When wiring any project, it's best to cross off each component as you wire it into the circuit. Check for poorly soldered connections. You should measure − 3 to − 5 volts at the collector of Q1 and Q2.

When operating inside the house, clip the antenna lead to a metal lamp base, screen, grounded pipe, or metal window frame for better local reception. For greater volume, connect the antenna wire to a metal ground or outside antenna. With broadcast stations thirty miles or so away, connect the flexible wire to an outside TV or radio antenna. You may add more solar cells for greater volume and distance.

PROJECT 3
SOLAR NOISE GENERATOR

Here we have a noise generator which may be used to inject a tone into a receiver or any other audio circuit when attempting to locate a defective stage. Simply, connect the ground clip to the chassis. Then place the probe tip on the base and collector of each transistor. A 2 kHz tone may be heard from the speaker if the stage is normal (Fig. 3-1).

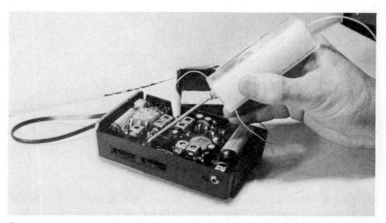

Fig. 3-1. This noise generator is signal tracing a radio chassis. Clip the ground lead to the chassis and place the probe tip at the base and collector of the suspected transistor.

Since the generator operates on 6 to 9 volts, an external solar power source is connected through a three foot cable. The Universal Solar-Pak (Project 16) is ideal to operate the noise generator. When operated from a 9 volt source, the generator pulls around 11 mA of current. The generator may be operated from 6 volts with a corresponding reduction in volume.

Wiring the Connector

The noise generator is constructed around a 555 IC with only four other components (Fig. 3-2). All parts are mounted on the IC

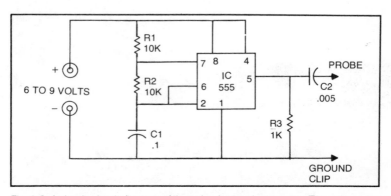

Fig. 3-2. A simple circuit diagram of the noise (or tone) generator. The unit can be wired in sixty minutes.

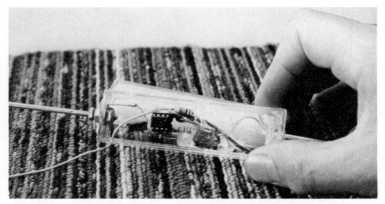

Fig. 3-3. Notice how the parts are laid out. Follow the schematic (Fig. 3-2) and Figs. 3-3 and 3-7 for parts layout and hookup.

socket terminals. The components are mounted as they are wired into the circuit. Of course, the signal output will depend upon the light striking the solar cells. The tone of the noise generator depends upon C1, R1, and R2. To raise the tone frequency use a smaller capacitor. For a lower frequency you can add more capacitance in parallel with C1.

Keep all component leads as short as possible. Mount the components as shown in Fig. 3-3. Place insulation over the connecting wires between pins 4 & 8 and 2 & 6. Keep the components close together so they will mount easily within the small plastic box (Fig. 3-4). The IC is not placed in the socket until all wiring has been checked and completed.

Fig. 3-4. This is a top view of the wired generator. Mount all components to the IC socket.

After all wiring is completed, go over the schematic diagram several times (Fig. 3-2). Mark or check off each component. Make sure they are tied to the correct terminal. Terminal number 1 of the IC socket corresponds with the dot on top of the IC. Bend down the socket tabs to help separate the soldered connections.

Choose a coupling capacitor with an operating voltage of 450 volts for C2 (Fig. 3-5). This prevents voltage breakdown when signal tracing circuits with high operating voltages. Not only does it protect the operator but it also may eliminate damage to the IC and other components. The ground return wire clipped to the chassis helps to prevent voltage surges and stray signal injection.

Plastic Box Construction

Practically any type of plastic box may be used to house the small noise generator. Here a discarded crystal cartridge box was used since the plastic is fairly thick and well constructed. The top lid was filled with a piece of foam to hold the components, while the IC and other parts are mounted in the bottom plastic case.

Before mounting any part, drill a ⅛ inch hole in the center of both ends of the plastic case. The front hole will be used for the metal probe tip and the rear hole for the solar power cable. A small side hole is drilled near the front of the case for the ground wire clip. All of these holes may be formed with the tip of a small soldering iron.

The probe tip may be constructed from a long 6/32 bolt (Fig. 3-6). Any long piece of metal with a threaded end will do. Simply grind a sharp point on one end. Place two nuts on the threaded end, then insert through the plastic hole. Place a solder lug on the inside

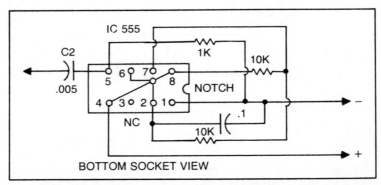

Fig. 3-5. This drawing of how the parts are connected. Observe the small notch in the IC socket for the correctly numbered terminals.

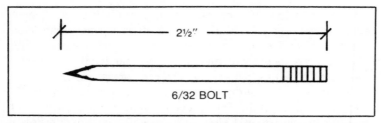

Fig. 3-6. The small metal probe may be constructed from a long 6/32 bolt with three nuts. Grind down the metal end so it can be easily touched to the different transistor terminals.

and use another nut to tighten the probe in place. After tightening the inside nut, snug the outside nut against the other so the metal probe will not become loose. Apply a coat of rubber silicone cement over the nut to hold it in place.

Now, lay the small IC with the other components inside the plastic case. Solder C2 to the small metal lug. Insert a two-wire cable in the bottom end and tie a knot so the cable will not pull out. Flexible speaker cable is ideal. Solder the cable to the ground (−) and probe (+) terminals shown in Fig. 3-7.

At the other cable end, connect two small banana plugs so that they can be inserted into the universal solar-pak. Place the red plug on the positive (+) lead and the black banana plug on the negative (−) lead. Double check the polarity by checking the negative lead with the ground clip. Your ohmmeter should show continuity when the two leads are touched. Now solder a small alligator clip to the

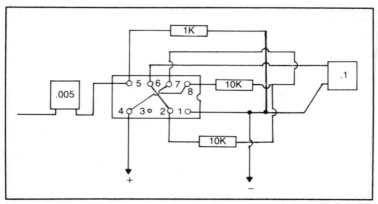

Fig. 3-7. Here is a pictorial layout of all parts and wiring. Mark off each component as you wire it into the circuit. Double check all connections before attempting to use the unit.

Table 3-1. Parts List for Project 3.

IC 555 (Radio Shack #276-1723 or equivalent)
IC socket (1) 8 pin (Radio Shack, #276-1995 or equivalent)
R1 & R2 10K ½ watt carbon resistor
R3 1K - ½ watt carbon resistor
C1 .1 of 100 volt coupling capacitor
C2 .005 of 450 volt ceramic capacitor
Banana plugs (2) (Radio-Shack #274-721 or equivalent)
Misc. 3 ft. light speaker cable, hookup wire, cartridge-type plastic box.

Construction time — 2 hours
Cost — $10 or under

ground lead at the front of the case. Insert the 555 IC with the tip dot over the pin 1 or the "V" indent towards pin 1.

Place three or four spots of clear silicone cement at various places around the components to hold them in position inside the plastic case. Check the unit before adding the top cover. If a separate audio amplifier is handy, connect the clip and metal probe to the input and listen for a tone in the speaker. Or you may check out the tone generator by removing the phono lead from the amplifier and inserting the metal tip of the generator. Another method is to actually use the noise generator in checking out a radio chassis.

Troubleshooting a Radio Chassis

Use a noise generator when the radio is weak or has defective stages. Start at the center terminal of the volume control and inject the signal from the generator. Turn the radio volume control wide open (Fig. 3-8). Connect the ground clip of the noise generator to the radio chassis. You should hear a loud tone in the radio speaker if the audio stages are functioning. If not, troubleshoot the audio stages with the noise generator.

Go to the base terminal of the 1st audio stage. If there is no signal, proceed to the collector terminal of the same transistor. Likewise, do the same with the driver and output transistors. You will notice lower volume as you proceed towards the speaker or output transistors. In fact, the signal may be very weak on the base terminal of the final output transistor.

When you hear the signal within the speaker, back up a step. It is likely that the trouble is between these two stages. For instance, if the tone can be heard on the collector terminal of a transistor and not on the base terminal of the same transistor, you may assume the

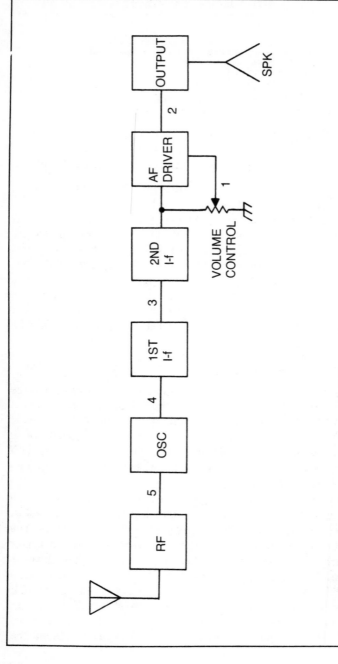

Fig. 3-8. All components and areas in which to inject the signal are indicated by numbers. Simply follow the numbers in checking the amp and front end of the radio.

transistor is defective. Likewise, if a signal is heard on one side of a coupling capacitor and not on the other side, the capacitor is open.

If a loud tone is heard when the generator probe is on the center terminal of the volume control, the audio section is normal. It is likely that the defective component is in the i-f, converter, or rf stages. Go to the collector and the base of each transistor until the injected tone ceases. The defective component is close at hand. You may notice that the signal in the rf and converter stages is quite weak. All types of transistor and tube radios may be signal traced with this solar noise generator.

Signal Tracing a Phonograph

A weak or dead phono amplifier may be checked with the noise generator. Simply disconnect the red and green wires from the crystal cartridge. Clip the ground lead to the black and white leads. Now touch the generator probe to the red and green leads. With the volume wide open, you should hear a loud tone in the speaker of each channel. If not, proceed to the 1st af stage. Place the probe on the base terminal of the 1st af stage. Be sure the volume control is wide open. Clip the ground lead of the generator to common ground or chassis. Proceed to the base and collector terminal of each stage until a signal is heard. When a signal is heard, back up a step and recheck. If no signal is heard with the probe connected to the final output transistor, you may assume the output transistor is open. Usually, you will hear a faint tone in the speaker at this point.

Signal Tracing the Stereo 8-Track and Cassette Player

You can check to see if the stereo tape head is open on a stereo 8-track or cassette player. Simply ground the clip to chassis. Turn the unit on. Clip a lead across the cartridge switch so that the capstan is rotating. Rotate the volume control wide open. Now, touch the generator probe to the ungrounded head connection. You should hear a loud hum in the speaker. No tone indicates a defective preamp or amplifier stage. Check each amplifier stage as in the phono or radio amp section.

A tone signal from the generator will quickly identify if the aux phono and tape jacks are normal in a stereo amplifier unit. Clip the ground lead to the chassis. Insert the probe into each jack, as the function switch is rotated. A loud tone from each jack indicates the amplifier section is normal. Check the cables from phono, tape, and aux for possible breakage or poor jack connections.

This is a great example of solar energy at work. Simply place the cells in the sun and watch the windmill rotate. The small windmill may be operated in the house under a 100 watt bulb. With a solar cell or two and a small motor and propeller you have most of the materials for the solar windmill. The whole project costs less than $25.00 to build (Fig. 4-1).

Forming the Tower

All tower material may be picked up at your local hobby or craft store. Hollow brass rods were chosen for easy soldering. Select two 36 inch × ⅛ inch diameter brass rods for the outside tower supports. After being formed, your tower will be about 18 inches in height. You will need two 1/16 inch brass rods, 36 inches long. The smaller rods are used for crisscross support material. See Table 4-1.

Mark the middle of the two larger rods with a felt pen. The joint will become the center at the top of the tower. Now bend these rods around a regular size spray paint can or equivalent diameter object. Keep the mark you made right in the center of the can. If you wish, both rods may be bent or formed at the same time. Take it slow and

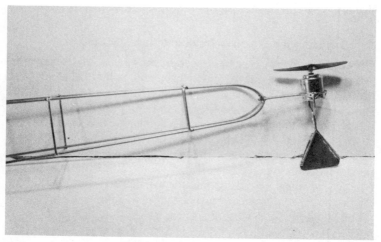

Fig. 4-1. This is the solar windmill ready to be mounted on a wooden base. The propeller should be cemented to the motor shaft after the windmill is completed.

Table 4-1. Parts List for Project 4.

Table 4-1. Parts List for Project 4.

```
1 —  low current solar cell motor #40,872
      Edmund Scientific Co., Solar cell motor
      Silicon Sensors, Inc., J4808 — GC Electronics
          #SP — 288                    — John Meshna, Inc.
      TM 21K896 — H & R Inc., #3212 Poly Paks, Inc.
Solar Cells  —  #3862 Poly Paks, Inc. #Q5541 — H & R Inc.
              #SP-287A — John Meshna, Inc. #42,741 — Edmund Scientific Co.
              #276-123 — Radio Shack
              #J4-804 — GC Electronics
Tower Material  —  Found at most Hobby Stores
              2 — ⅛ inch brass rods — 36 inches long
              2 — 1/16 inch brass rods — 36 inches long
Misc. Hookup wire, wood base and solder

Construction Cost — Under $25.00
Construction Time — 8 to 10 hrs.
```

careful. You do not want to pinch the rods at the center and top. These brass rods are easily bent and formed.

Make up a couple of simple forms to hold the tower together and in line while soldering up the brace bars. Lay out a three inch square on a heavy piece of cardboard. Inside this square, draw a 2 inch square (Fig. 4-2). Use a ruler or T square to make the squares accurate. Double check by measuring from corner to corner. Punch

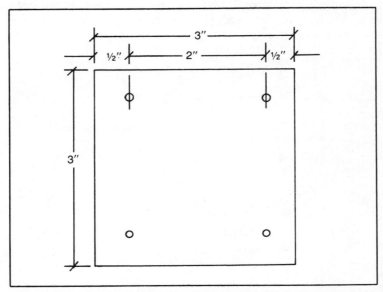

Fig. 4-2. The layout of a two inch square to hold the top side of the tower supports in position. Use heavy cardboard for these forms.

the holes with an icepick or nail in each corner of the inside of the two inch square. This will become the top side support form.

Now, make up a four inch square piece of heavy cardboard. Inside, draw a three inch square. Check the inside for perfectly square corners. Punch a hole in each corner. Do not make these holes too big since they should fit snugly on the brass rod support legs (Fig. 4-3). This three inch cardboard square will be the bottom support form.

Slip the three inch square onto the tower. Place each leg into the opposite corner of the two inch square cardboard holes. Place the three inch cardboard four inches down from the top of the tower. Now, push the four inch piece of cardboard up about two inches from the bottom. Slide a short piece of 2 × 4 wood underneath the cardboard, between the tower legs. Push the cardboard down tight against the wood piece. Straighten the tower so it's plumb up and down.

Installing the Support Rods

After the two cardboard pieces are in place, cut four rods of the 1/16 inch material, three inches long. These will be the bottom support braces. When cutting this 1/16 inch rod, use a large pair of side cutters. Make sure that both ends of the rod are cut the same way. The flat end area created by the sidecutter will lie flat against the tower support rods.

Fig. 4-3. Fit the four inch cardboard form over the bottom part of the tower to hold the legs together. Solder the 4 inch tie bars in place.

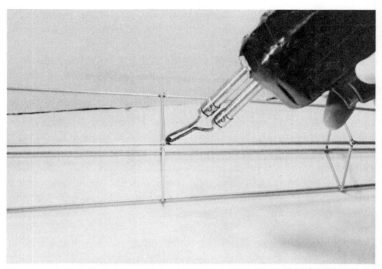

Fig. 4-4. Solder the tower legs with a large soldering iron. The brass legs and supports do not need any soldering flux. Rosin core solder paste may change the color of the brass rods at the soldered connection.

Use a large soldering iron (150 to 300 watts) and tin the area next to the cardboard. Solder the two top loops of the tower together. Make sure the cardboard is down tight against the 2 × 4. Check to see if the tower is perpendicular. Now, tin the flat side of each end of the 3 inch support rods. Lay the tinned side towards the tower. Hold the rod in place with a pocket knife or pencil. Solder each corner with rosin core solder. You do not have to tin these brass rods with soldering paste (Fig. 4-4).

Solder each bottom support rod into position. After they are tacked in, lay the tower on it's side and carefully apply more solder to the corner joint. Make the solder joint round, hiding any rod edges. After making a few corners, a neat job can be obtained. Hold the rod tight against the tower rod when making these round edges.

The top cardboard may be held in position by leveling up the cardboard. Place a layer of masking tape under each tower rod to hold the cardboard in place. Now, take the 2 inch rods and solder against the cardboard area. Always tin the tower rods before placing the support rods in position. Make all corners round with solder.

Divide the distance between the top and bottom support rods. In this case, the distance was 6¼ inches. Mark each tower leg 6¼ inches from the bottom support rods, so that these rods will be level. Cut four 2½ inch rods for the center of the tower. Tin each area. Lay the tower on it's side and start to mount the center rods. Hold the tower legs together so the support rods can be soldered together.

After selecting a 1.5 volt motor and propeller, form a loop around the motor with a piece of 1/16 inch rod. Cut the rod 7 inches long. Carefully bend it around the motor housing. Form a tight fit and bend the rod at right angles (Fig. 4-5). The bottom support rod should stick down 2¼ inches so the motor will mount 2 inches above the tower.

Fasten the rod mount to the motor with three small areas of the motor bell. Most motors have a brass or metal housing. Tin the area and use soldering paste, if needed. Be careful not to melt the plastic of the motor frame. If the motor shell cannot be soldered to the support rod, cement with silicone rubber cement. Allow the support to dry over night.

The motor frame should be soldered to the top of the tower. Hold the frame straight and tack one side at a time. Fill in the gaps with solder. Round up both sides. If the motor frame is out of line, loosen up the solder on one side and try again. The small framework can be straightened very easily. Use a knife or round rat-tail file to trim up the soldered areas.

The tail of the windmill can be constructed of two pieces of 1/16 inch rods bent to fit around the middle and out the back. Bend the ends away from each other. The fins or tail may be enclosed with one solid piece or use flat brass shim material. If a piece of flat copper is handy, use it. Straighten and bend the tail assembly so it will fit around the motor frame. Solder the tail assembly into position. After mounting, bend the tail assembly in line with the motor assembly. Do not cement the propeller to the motor shaft until the whole unit is mounted and wired.

Mounting the Tower

Select a pine or oak board to mount the tower on. Cut the board at least 6 × 7 inches. If a router is handy, curve the board edges. Sand down all sides of the mounting board. A ready made plaque found in the hobby stores will work as well. In fact, these boards are already sanded and need only a coat of paint or stain.

Before applying the final finish, drill all four tower mounting holes. Place the tower in the center of the mounting board. Mark each leg. Drill the holes slightly larger than the ⅛ inch diameter tower leg material. Drill completely through the wood. Now, place all four tower legs into the drilled holes. You may have to ream out

Fig. 4-5. A closeup view of the motor support frame. The small brass rod may be formed around the motor shell and cut, before bonding to the motor itself.

the holes some, with the drill, to get the legs to fit.

Temporarily lay the two solar cells on the tower base. The cells will stick outside the tower area. Drill two small holes so the solar cell wires will fit inside them and connect underneath the tower. If one wire from the motor is needed to connect the solar cells, drill a small hole alongside the lower leg at the back of the tower. Use very fine sandpaper and go over the entire top surface.

When using oak or other hardwood for the tower base, the board may be finished with a clear spray or lacquer. Whatever finish used, at least three coats should be applied with ample drying time allowed between coats. Lightly sand the board between each coat. The mounting board should be thoroughly dry and finished before mounting the tower.

Wiring the Tower

One leg of the tower will be used to carry power to the motor. Ground one lead of the motor to the tower frame. The other small motor wire may be cemented to one inside leg of the tower. Drill a small hole in the hollow log of the tower so the small power wire can be hidden. Be careful not to drill through the tower leg and weaken it. Ream out the small hole so the edge will not cut into the wire insulation. Also, you may simply wrap the wire around one tower leg.

Start the small insulated wire at the top and feed down through the metal tubing. Pull out at least six inches at the bottom of the tower. Form the small wire at the top around the motor support and solder to the remaining motor lug. Mount the tower to the finished base and cement with clear silicone rubber cement.

One solar cell is sufficient to operate the small motor if a 100 watt bulb is to be used above it. The bulb must be placed within a foot of the tower base. For sunlight operation or less light, connect two solar cells in series.

To connect the solar cells, feed the leads down through the holes in the wooden base. Connect the two cells in series. Solder the red wire to the black wire of the next cell. The remaining red and black wire will go to two wires connected to the motor (Fig. 4-6). If the motor runs backwards, reverse the two motor connecting wires at the base of the tower.

Checking the Wiring

Double check the complete solar cell to motor wiring. Before the solar cells are cemented into position make certain that the motor will rotate. Place a 100 watt bulb close to the solar cell. The motor should start to operate. If not, rotate the motor shaft by hand to get it started. Check the voltage at the motor terminals if the motor will not rotate.

Set the vom at the 3 volt scale. Hold the lamp bulb close to the solar cells. You should measure 0.5 volt with one cell and 1 volt with two solar cells in series. If the correct voltage is found at the motor and it does not rotate, the motor is defective (Fig. 4-7). Check the

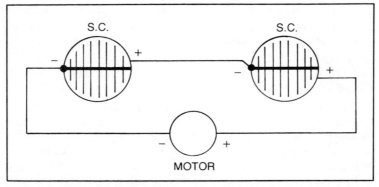

Fig. 4-6. Connect the two solar cells and motor in series. The solar cells should be at least 0.5 amp or 1 amp cells. Select a motor with operating current under 300 mA.

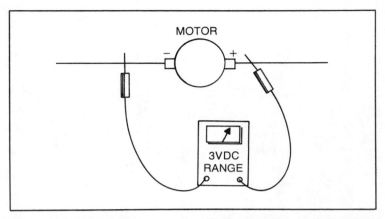

Fig. 4-7. When the motor will not rotate check the voltage at the motor terminals. No voltage at the motor terminals may indicate a poorly soldered connection or improper tower ground.

wiring of solar cells and tower wiring when no voltage is found at the motor terminals. Inspect the connecting wire placed inside the lower leg for a possible ground to the metal framework.

Although the solar windmill is nothing more than a novelty item, it's still fun to build. Simply set it by your window and watch it spin. It requires no batteries, no fuel, and no repairs. Here is a great example of solar energy at work.

PROJECT 5
SOLAR CODE OSCILLATOR

Here is a simple to construct code practice oscillator powered by a solar cell. You may practice code at anytime or any place, without worrying about battery replacement. In fact, this code practice oscillator can have a lifetime shelf-life without any problem. You can practice morse code under a reading light or in the sunshine (Fig. 5-1).

Only four separate components are used except for the external practice key. The code practice oscillator is constructed around a Prezo tone buzzer. This small buzzer will operate on 1.5 volts with less than 12 mA of current, and produce adequate sound. Greater volume may be obtained by increasing the applied dc voltage.

A 1.5 volt dc solar cell was used to power the buzzer unit. This voltage may be obtained by selecting a solar unit with a 1.5 volt

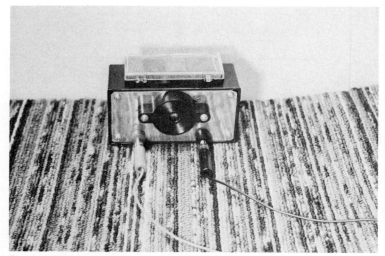

Fig. 5-1. This is a photo of the completed project.

rating or three small cells placed in series. Since the operating current is very small, almost any solar cells will be sufficient. Simply wire them in series (Fig. 5-2). See Table 5-1.

Preparing the Front Panel

Select a project plastic case or equivalent to mount the various components. The front panel may be constructed from plastic, metal, or wood. If a piece of metal is used, the small key jacks must be insulated from the metal panel. A plastic front panel is ideal since no insulation is needed. A metal front panel was used in my project.

All components except the solar cell are mounted on the front panel. Center the buzzer or indicator on the metal piece. Keep the

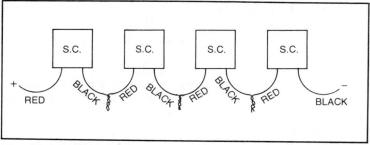

Fig. 5-2. Four small crescent solar cells are wired in series to operate the code practice oscillator.

Table 5-1. Parts List for Project 5.

BL — Indicator — Prezo Buzzer — Radio — Shock 273-060 or
    equivalent   ($2.99)
SC — Solar Cell or Cells — GC Electronics J4-803 (1.4V)   ($9.45)
    Microgenerator Cell — Edmund Scientific Co. 42 —465 (1.5V)
    or make up your own series cells.   ($6.95)
J1 & J2 — S122 — 125 MA — 4 for $6.00 Solar Amp Inc.
Phone jacks — Radio Shack banana jacks 274-725 or equivalent.
Code Produce Code Key — Radio Shack 20-1084 ($5.95)
    or GC Electronics Telegraph key 24-800 ($2.65)
    or equivalent. Gravois Merchandisers Inc. #9A3093-3.
Project Case — Radio Shack 270-222 ($1.99) 4¾″ × 2½″ × 1 2/5″
    or GC Electronics H4 — 722 — 5″ × 2½″ × 1⅝″ ($2.85)
Misc. Flexible key hookup wire, 2⅛″ bolts, nuts, solder,
    silicone rubber cement.
Cost of Project — $12 to $15.00 less code practice key.
Construction time — 2 to 4 hrs.

top of the buzzer even with edge of the panel. Drill two ⅛ inch mounting holes. Drill two ¼ inch mounting holes on each side of the indicator for the two banana jacks. Check Fig. 5-3 for front panel dimensions.

After all holes are drilled, mount the three components. Be sure and insulate the key jacks when a metal front plate is used. Either wrap a small ¼ inch piece of plastic tape around the metal jack area or drill larger mounting holes so the fiber mounting washer insulates the jacks from the metal panel. To make certain that the metal jack area is not touching, check each jack contact for a possible ground to the metal front piece with an ohmmeter.

Preparing the Cabinet

Only two 1/16 inch holes are needed at one end of the solar cell for proper hookup. The solar cell may be glued with silicone rubber

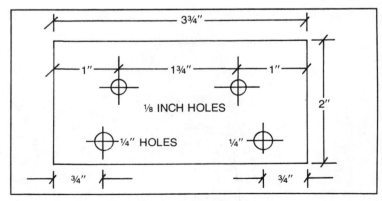

Fig. 5-3. The front panel holes and panel dimensions.

cement to the plastic case. First, poke the red and black leads down inside the plastic case and place a drop of cement on each end of the cell area. Some solar cells come in a small plastic case. To protect the cell, leave the solar cell in the plastic case and cement it to the top of the project area.

Center the cell before the cement sets up. Place a book or other flat object on the cell to hold it in place. Let the project set up over night for proper bonding. The plastic case protects the solar cell from collecting dust.

If three separate cells are to be used for solar power, connect the cells in series before mounting. Temporarily lay the cells on the top of the case and solder them together. Connect a red and black lead together at each connecting cell (Fig. 5-4). The top side of one cell should be connected to the bottom side of the next cell. When finished, you should have one black and red lead.

To connect solar cells without connecting leads, solder one lead to the top of the cell and another lead to the bottom of the same solar cell. Use very fine copper wire for these connections. Regular hookup wire may damage or break the cells.

The top side of these cells are the negative (black) connections. Scrape the silver area until a shiny surface appears. Be careful not to damage the cell. Use a low voltage soldering iron and tin a spot to connect the small copper wire. Likewise, do the same to the bottom side of the cell (positive).

Wiring the Project

Run the two buzzer wires through the hole in the front panel and connect the black wire to the black key jack. Solder the red wire

Fig. 5-4. Mounting the separate cells in series on the top panel of the plastic case. Connect a black wire to the red wire of the next cell.

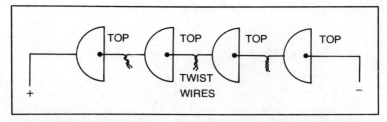

Fig. 5-5. The separate crescent solar cells are wired in series. Connect the top side of one cell to the bottom side of the next cell.

from the solar cell (positive) to the red key jack. Connect the red buzzer wire to the black solar cell lead.

Actually, all components are wired in series. Color coding of component wires may be followed, but is not required. If each component is wired in series, the code practice oscillator will function (Fig. 5-5). The solar cell polarity may be observed without any problems. If in doubt, follow the pictorial diagram shown in Fig. 5-6.

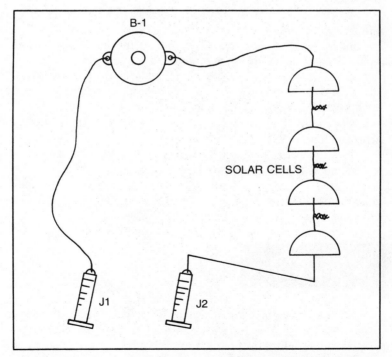

Fig. 5-6. This is a pictorial diagram of how to wire the code practice oscillator.

Fig. 5-7. Check the wiring of each component. Use an ohmmeter if you have one.

Testing

The code practice oscillator can be checked by simply shorting the two front jacks together. You should hear a 4 kHz tone. The volume may be raised or lowered by moving the unit near a reading light. Greater sound may occur under direct sunlight.

If a tone is not heard, bring the light closer to the solar cell. A low or no sound indicates poorly soldering connections. Recheck each connection (Fig. 5-7). Sometimes when the front cover is replaced, a wire may break off. If handy, check the wire continuity with an ohmmeter.

Now, insert the code key and tap away. The key can be left connected as the circuit is not completed until the key is pressed. For greater volume connect more solar cells in series. This code practice oscillator may be operated under a reading lamp if 4 to 6 cells are connected in series.

PROJECT 6
SOLAR DIODE TESTER

Here is a solar diode tester which is fun to build and provides a useful service for the electronics enthusiast. You may check suspected silicon diodes in or out of the circuit. If you have a handful of those unmarked surplus diodes around, you can test them for quality

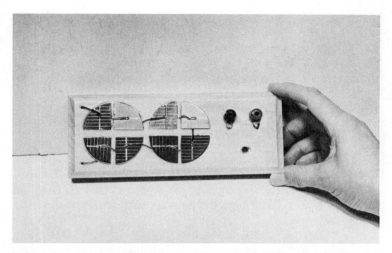

Fig. 6-1. This is a top view of the solar diode tester showing the solar cell layout. The solar tester may be constructed inside a plastic box if desired.

and even mark the cathode terminal. All of this is accomplished with several solar cells and a light emitting diode (Fig. 6-1). Practically any small LED will work. Check for one with a low operating voltage and current rating. The voltage should be somewhere between 1.5 and 5 volts with a current rating around 20 to 50 mA. These light emitting diodes come in jumbo and miniature size, red, yellow, and green colors. The red color was chosen for this project. See Table 6-1.

Table 6-1. Parts List for Project 6.

8 — Solar cells may be pieces, chips, crescent or ¼ round type cells.
Quarter round solar cells:
    # TM 21K666 H & R, Inc.
    # S122 — Solar Amp, Inc.
    # H-11 — John L. Meshna, Inc.
    # 42,268 — Edmund Scientific Co.
LED — Red light emitting diode
    #276 026 — Radio Shack
    #360VA — E.T. Co. Electronics
    #H-49A — John Meshna, Inc.
    #J4-940 — GC Electronics
Misc. — Red and black banana jacks, 3″ × 8″ mounting doors (found in hobby shops), hookup wire, solder, etc.

Cost of Project — $15 to $20.00

There are no switches used in the project since the circuit is not completed until a diode is inserted in the test jacks or light is provided above the solar cells. In fact, there are no batteries to wear out or replace. The little solar diode checker can be used for years without any maintenance.

Selecting the Solar Cells

When a 20 mA LED is used for the indicating device, any size or shape solar cells may be used to power the tester. You may want to choose 25 mA crescent or solar cell chips for this project. Here is a chance to use up some of those broken pieces of solar cells. If a LED with larger operating current is used, choose a cell with 75 mA of current or higher. They may be half round, quarter, or crescent type solar cells.

A 20 mA 1.8 V LED was used in this solar diode tester (Fig. 6-2). Only 8 solar cells are needed for adequate operating voltage. Under strong sunlight the total output voltage from the solar cells may be greater than 3.5 dc volts, while under a reading lamp, the tester works nicely with 2 volts or less. If a LED is used with a working voltage of 3 volts, connect ten solar cells in series. A 5 volt LED may be operated from thirteen solar cells connected in series.

Preparing the Board

Before mounting any solar cells, prepare the mounting board. Lay out the holes to be drilled by taping a piece of cardboard or paper to the top of the board. If the board comes wrapped in cellophane, leave it on and mark holes with a felt pin. Not only will it keep the board clean but provides a layout for the holes to be drilled (Fig. 6-3).

Drill two very small holes in the left side corners for solar cell connecting wires. Two 5/16 inch holes are drilled in the top right

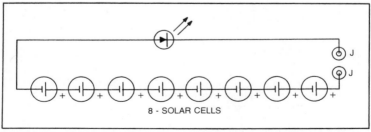

Fig. 6-2. The eight solar cells, LED, and test jacks are all wired in series. The operating voltage and current will determine how many solar cells are needed.

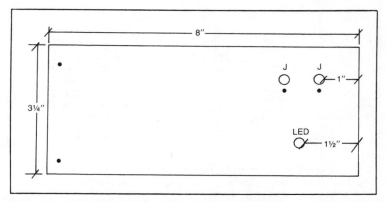

Fig. 6-3. Layout the holes to be drilled on a piece of cardboard and use masking tape to hold it in place. Try to keep the pieces of board clean so a clear finished appearance may be obtained.

section to mount the banana test jacks. Under each banana jack, drill a small hole to connect the jacks into the circuit. Use the same drill for mounting the small LED. Drill the LED indicator hole at an angle so the overhead light will not hit the surface of the LED.

After the holes are drilled, remove the cellophane or cardboard from the mounting surface. Use a very fine grade of sandpaper and dress down the top and bottom sides. Go over the end areas with sandpaper. Lightly touch up all edges. If a beveled area is found, lightly sand down these edges. When wiring and cementing the various components, keep the surface as clean as possible.

Mounting the Solar Cells

Before mounting the solar cells, solder a short piece of connecting wire to the bottom side of the cell. After all cells have a connecting wire soldered to them, cement the cells to the board surface with clear silicone cement. If the connecting wire is fairly large and makes the cell raise above the surface, cut out an area underneath each cell so it will lie flat. A small indentation with a pocket knife will do.

The cells may be mounted in any configuration or line-up (Fig. 6-4). When ¼ inch cells are used, line them up in a round pattern. If small pieces of solar cells are used, they may be lined up by drawing a pencil line down the board area. Apply a dab of cement to the back of each cell. Slide the cells around for the best appearance.

Temporarily line up the cells where they may mount. Then solder the connecting wires to the backside for easy hookup. Usu-

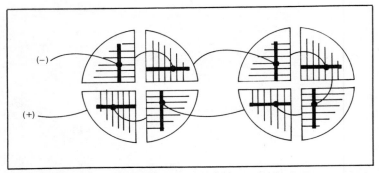

Fig. 6-4. When quarter round solar cells are used, line them up in a round cell type pattern. All solar cells are connected in series to obtain the operating voltage.

ally, if the wire is soldered to the center of each cell and to the edge of the cell, they may be connected rather easily. Of course, one cell may be soldered at the flat edge and the next cell towards the rounded edge. The wire for the fourth cell should be soldered to the bottom flat edge so it will connect to the cell below it.

Place the black banana jack in the left hole and the red jack in the right hole. Cement them into position with rubber silicone glue. Mount the small LED in the bottom hole. Place the indicator inside the hole about ¼ inch from the top surface. The LED is placed in this position so the top section faces the operator. Cement the LED into position with the leads flattened back against the bottom side of the board.

Connecting the Cells

Start at the top left cell and connect to the long black wire to the top side of the cell. The other end of the black wire will go to the black banana plug jack. Pull the wire tight and solder at the jack connection. Now, connect each wire to the top of the next cell. All cells are wired in series.

Cut the wire just long enough to tie to the center bar area of the cell. Tin the wire and solder in place. Use the blade of a pocket knife or lead pencil to hold the wire on the bar connection. Do not apply too much heat or you may lift the solder from the cell. Connect the fourth cell wire to the bottom of cell five. The long red wire solders to the back side of the board and is soldered to the cathode terminal of the LED.

Check Fig. 6-5 for correct wiring terminals of the LED. The anode (−) terminal connects to the red wire (−) from the solar cell.

Now, connect the cathode terminal to the red banana jack, keep the wires underneath the board as tight as possible. These two long wires may be held flat against the board with a couple of dabs of silicone rubber cement. Lay the long nose pliers on top of the wires to hold them in position while the cement dries.

Checking the Diode Tester

After cement has set-up and all connections are soldered, place a 75 watt bulb or lamp over the solar cells. The small LED should light up with the red and black banana jacks shorted together. Take a short piece of wire or test lead and place down inside the jack. Look straight down on top of the LED. You should see a red glow. If not, use a vom for voltage tests.

Measure the voltage at the black and red terminal wires of the solar cells. Under artificial light, you should measure a little over 2 volts. If placed in the daylight over 3.5 volts is measured. Check the soldered connections of each cell if the voltage is less than 2 volts. When voltage is above 2 volts, and there is no LED indication, reverse the two wires going to the LED. Sometimes it's difficult to see the small flat side of the emitting diode for identification. Now check for the red light of the emitting diode by shorting the two banana jacks together.

The suspected diode may be checked for open, shorted, or normal conditions, by placing the diode wire ends down into the banana jack terminals or with plug-in test leads (Fig. 6-6). Diodes outside the circuit are easily tested by slipping the ends inside the jack area. If the diodes are located within the circuit, use test leads and tips for testing the diodes.

A normal diode will light in one direction and not with reversed leads. With the cathode terminal of the diode connected to the black jack and the other bare lead to the red jack, a normal diode will light up the LED. Now, reverse the diode. If the light comes on, the diode

Fig. 6-5. This is the bottom view of a red light emitting diode. Connect the red wire (−) of the solar cell to the anode terminal of the LED.

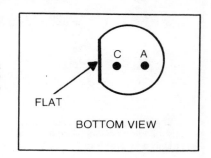

FLAT

BOTTOM VIEW

is leaky or shorted. A normal diode will light in only one direction. No light from the LED indicates the diode is open in any direction.

The brilliance of the LED will be determined by how much light strikes the surface of the solar cells. This LED lights from 1.8 to 2.3 V at 20 mA. Under sunny or even cloudy days, the tester functions very well. While under a fluorescent shop light, you may not receive enough light to operate the tester. While under a desk or reading lamp the results are great.

Fixed diodes come in many sizes and shapes. Usually, the cathode terminals are marked with a silver ring at the end or have other identifying marks. If you have some of those surplus silicon diodes without markings, you can determine the cathode terminal (+) with the solar diode tester. Place the diode ends into the banana jacks. When the LED lights up, mark the terminal end going to the black banana jack with a drop of white paint. Liquid paper correction fluid does a nice job. Check and mark all unmarked diodes in this manner. Don't forget to check them for leaky or shorted conditions.

Finishing Touches

The top of the solar cells may be covered with a sheet of plastic or a plastic box to protect the cells. Several coats of liquid plastic may also be applied over the entire top and end section of the solar diode tester. Four coats of valspar polyurethane liquid plastic was applied to the finished project. Follow the manufacturer's directions

Fig. 6-6. A suspected diode may be checked for open, leaky, or normal conditions by inserting the diode into the banana jacks. A normal diode will only light the LED in one direction.

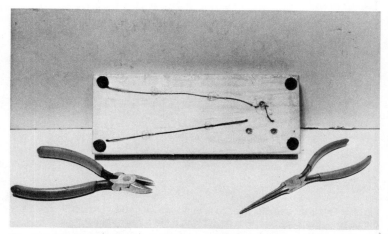

Fig. 6-7. For the finishing touches, file or cut off any excess plastic and place rubber feet in each corner. Cement rubber grommets if rubber feet are not available.

of application on the paint can. This polyurethane liquid plastic may be purchased at most paint, hardware, or discount stores in practically any size can.

Brush the liquid plastic underneath any cell raised above the board surface. This will prevent breakage if the tester is dropped. Now, apply heavy coats of liquid plastic on top of the tester and brush out any bubbles. Keep the liquid out of the banana jack holes. Let the liquid drop off the ends into a cardboard box. Apply the liquid plastic outside and away from all flames. Each coat of plastic may take a couple of days to dry before another coat is applied. For the finishing touch, cement four rubber feet or grommets to each corner of the bottom side of the tester (Fig. 6-7).

PROJECT 7
SOLAR AIRPLANE

In this project you can build a tower to show off your favorite model airplane kit and even have the propeller powered by solar energy. The tower may be constructed as in Project 4 or you may make a slightly different version. Simply mount a small low voltage motor behind the propeller. At the base of the tower mount one or two solar cells to power the small motor. Not only does the model airplane have a pleasing appearance, but here we see solar energy in action (Fig. 7-1).

Fig. 7-1. This is the completed project. Simply place a 100 watt bulb nearby or set the base of the tower in the sun and watch the small airplane fly!

Tower Construction

The tower may be constructed as in Project 4. Select two ⅛ inch hollow brass rods 36 inches long. Bend the brass rods around a paint can or any round object. Be careful. Although these hollow brass rods bend rather easily, try not to kink them. After the rods are bent in the middle, cut each leg to 14 inches in length. See Table 7-1.

Table 7-1. Parts List for Project 7.

Tower Material:
2 — Brass ⅛″ hollow rods — 36 inches long
3 — Brass 1/16″ hollow rods — 36 inches long
1 — Small motor — #140,872 Edmund Scientific Co.
                            Solar Cell motor. Silicon Sensor, Inc.
                            J4-808 GC Electronic
                            #SP-288 John Meshna, Inc.
                            TM 21K896 H & R, Inc.
                            #3212 Poly Pak, Inc.
1 — Mounting board 4×5 inches — May be plaque board found
       at Hobby and Art Craft Stores.
2 — Solar Cells .500 to 1 AMP at .5001 tS
       #3862 Poly Paks, Inc.
       #Q5541 H & R, Inc.
       #SP-287A John Meshna, Inc.
       #42,741 Edmund Scientific Co.
       #276-123 Radio Shack — 24-814 GC Electronics.
Misc. — Hookup wire and solder
Construction time — 10 to 14 hours.
Construction Cost, under $25.00 — with the solar cell under $35.00

Solder the two top sections in a crisscross pattern right in the center of the rods. Now, place a two inch square cardboard separator chart four inches down from the top. Place a three inch square cardboard separator at least two inches up from the bottom tower legs (Fig. 7-2). These same type of cardboard separators are used in the solar windmill project.

Cut four 2 inch lengths of 1/16" brass rod for the cross support braces. These are soldered into position at the top of the two inch cardboard separator. Likewise, cut four 1/16" pieces for the bottom supports. With both cardboard separators in place, the tower is held rigid and it is a lot easier to solder the brace supports. Now, cut two 1/16" brass rod supports to tie the two different brace sections together on each side of the tower. These crisscross supports should be individually cut and soldered into position.

You may want to build the tower like the old windmill tower. Instead of bending the ⅛ inch support rods, simply cut four pieces, 14 inches in length. At the top of the tower, form a 1¼ inch square with the 1/16 inch rods. Bend the small rods with a regular pair of pliers. It may take several attempts to square up the top brace support (Fig. 7-3). Leave the cut end joint in the center of the backside of the tower. Solder the 1¼ inch square at the top leg of the tower.

Cut a sheet of brass or copper scrap material to go over the top tower area. Overlap the metal material ¼ inch below the top brace

Fig. 7-2. The tower legs are held in place with a cardboard separator. Solder the bottom brace rods into position around the tower.

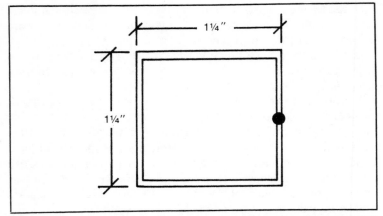

Fig. 7-3. Form a top brace support 1¼ inches square. The tower support legs will solder to each support corner.

supports. Now the top airplane support form may be soldered directly to the top flat of the tower end section.

Solder all tie support rods at each end of the tower legs. Round off the solder. You may apply more solder than normal and round off with a small file. Clean up all soldered joints with a small rat-tail file. Be careful not to scratch or mark the copper rod surface with the file or soldering iron.

Preparing the Mounting Board

Select a 4 × 5 inch (or larger) white pine board to mount the tower on. These plaque or decoupage boards can be found at most craft or hobby stores (Fig. 7-4). You can make your own, if you desire. Any form will do. Simply set the tower on the mounting board.

Center the tower and draw circles around the bottom legs. Some plaque boards come wrapped in cellophane. If the cellophane covers the board, leave it on until all holes are drilled. Mark the cellophane with a marking pencil or felt-tip pen. Use a ¼ inch drill bit and drill each leg hole completely through the board. Drill four small holes for the solar cell wires in the center of the board. If you like, drill one larger hole so all wires can be pulled through it.

Turn the board over and cut a V-shaped groove between the back two tower legs. Use a pocket knife or razor blade to make these cuts. Now, cut a line from the center holes to the center cut of the tower legs. This will form a "T". All of the connecting solar cells and motor wires will lie in these groove areas.

Although the board may be quite smooth, sand down all sides, and the bottom and top area with very fine sandpaper. Fill all broken and dented areas with plastic wood. Make sure all curved areas are properly sanded. Always sand with the grain of the wood. Remove any pencil or pen marks from the mounting board.

Apply a coat of stain over the mounting board. Also stain the bottom side. Wipe off all excess stain. You may use a wiping stain found at most hardware stores. For added appearance, apply gold leaf in the curved areas of the mounted base.

Let the mounting board dry for a couple of days. Either apply tung oil finish or spray on several coats of clear lacquer. Keep the mounting board in the garage or outside while spraying paint or lacquer. Sand lightly between each coat.

Before mounting the tower to the wood base, prepare the top of one back leg for a return connecting wire. Drill a small hole in one back leg at the top of the tower to insert a small insulated hookup wire. If the regular windmill tower is constructed, drill a small hole in the top sheet metal over one of the back leg supports. The connecting wire may be wrapped around one leg and down the tower.

Cut a piece of solid insulated hookup wire twenty inches long and insert it through the leg area. Leave about two inches at the top and four inches at the bottom of the tower. It's much easier to prepare and insert the motor wire before the tower is mounted. The motor connecting wire may be taped or glued to the outside of one rear tower leg. However, hiding it not only gives a better appearance, but also adds mystery to its mode of operation.

Motor Mounting

Select a very small solar motor to go inside the model airplane. The motor should be mounted when the kit is being glued together or the front part of the airplane may have to be removed. If the kit is all ready assembled, remove the front airplane motor assembly, mount the motor and glue the front piece back together.

The motor should be level and straight with the body of the plane. Otherwise, the propeller will be rotating at a different angle. Use small pieces of foam or wood to hold the motor level. Cement the motor into position with rubber silicone cement. Use clear silicone cement for this purpose. Colored or black cement may stain or discolor the body of the airplane.

It's much easier to mount the motor in only one body section. If you are putting the kit together, mount the motor in one half of the

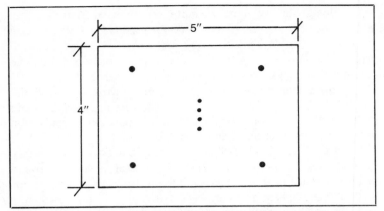

Fig. 7-4. A mounting board may be cut from oak or pine — 4 × 5 inches. You may want to use a decoupage or plaque board found at most hobby and craft stores.

body section. Level and square the motor. Now, let the cement set up overnight before assembling the rest of the plane. Usually, the motor leads are long enough to go out the middle of the plane's area. If not, connect flexible hookup wire to the motor terminals. The motor leads should be long enough to go out the center bottom area of the plane and connect to the existing motor wire and one metal tower leg.

The motor mounting is a little more difficult on a 7 inch model airplane. Often, the motor area is quite small. A *very* small solar motor must be used in these models. A ¾ inch diameter motor is a tight fit. The front body area may have to be reamed out with a pocketknife. This is no problem since plastic models are easily cut with a pocketknife.

Ream or cut out the front body area of both pieces. Make sure the two body pieces will fit together around the motor. Keep cutting away until it does fit. Recheck the motor mounting (Fig. 7-5). The motor may stick out quite a ways to clear the belt propeller assembly. Temporarily build the motor and plane body up to the propeller assembly area. Mark it and cement the motor into position.

Mounting the Tower

After the airplane kit and motor have been finished, the tower should be cemented to the bottom base board before the model is mounted to the tower. Mount the airplane last. Feed the connecting wire through the correct hole in the mounting base. Now, fit all four legs in their respective mounting holes and push the tower down into the base area. Let the tower legs go through the small base area. The other rear leg will have the motor ground wire soldered to it. Place a piece of hookup wire through one of the legs and solder it. This wire will connect the return wire to the solar cells and motor.

Apply rubber silicone cement in each bottom leg hole to hold the legs in position. Pull the tower up even with the bottom base area. Level and straighten the tower. Keep a piece of paper under the mounting base to collect the dripping cement. Place a couple of pencils or wood pieces under the base until the cement sets up.

Mounting the Solar Cells

Mount the solar cell in the center of the tower base when only one solar cell is used. Two solar cells may be mounted side-by-side. Keep them inside the tower area for protection (Fig. 7-6). Remember, a 100 watt bulb must be placed quite close to one solar cell to make the prop rotate; while two cells are more efficient.

Place a coat of clear silicone cement on the bottom area of each solar cell. Feed the wires from the solar cells down through the base area holes. If the cells will not lie flat, clean out the area with a pocketknife. Lay the cells down very carefully. Do not press down too hard or you may crack the cells. Wipe off any excess cement from around the cells and mounting area. Both the tower and solar cells may be mounted at the same time. Let the cement set up overnight.

Fig. 7-5. The motor should be placed inside the model airplane when assembling the kit. If not, cut out the front area so the motor shaft can stick through.

Fig. 7-6. Mount the two solar cells inside the tower area for protection. Run the leads through the base mount before cementing them into position.

Connecting the Cells

The solar cells and motor must be connected in series (Fig. 7-7). If the solar cells have one black and one red connecting wire, it's much easier to wire together. Start with one of the black solar cell wires and solder to the tower ground wire. Now, solder the other red wire to the remaining black wire of the second cell. The red wire of the second cell must connect to the motor wire which feeds through the tower leg. If the cells do not have leads soldered to them, place a red on top and black on the bottom of each cell.

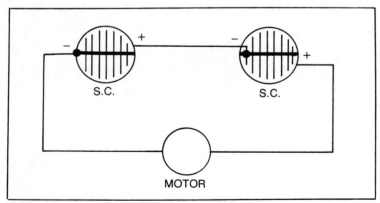

Fig. 7-7. Connect the solar cells and motor in series. Keep the leads as short as possible. After the motor is running, apply silicone rubber cement over the exposed wires under the mounting base.

When the model airplane is cemented to the tower mount, the red wire from the motor should connect to the wire through the tower leg. Ground the black or remaining wire to the brass tower. If the propeller is rotating backwards, simply reverse the two motor leads. Keep the leads from the airplane model close together and out of sight.

All solar cell leads must be cut as short as possible. Keep the leads within the cut "T" area. The wires may be held in the area with scotch tape. It's best to apply a coat of clear silicone cement over all of the wires to hold them in place. Make certain that the propeller is rotating before cement is applied.

Model Support Mount

The model airplane support must hold the plane above the metal tower. Of course, the length of the support rods will be determined by the size of the airplane. A cradle type support may be

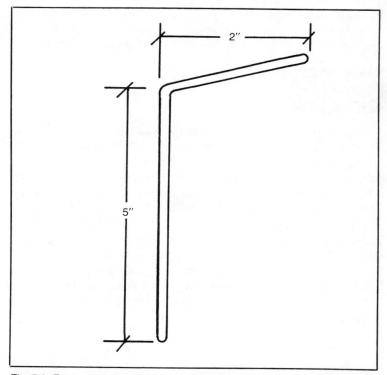

Fig. 7-8. Form a simple loop out of a 14 inch length of 1/16 inch tubing for the plane support. Tilt the nose upward.

used where the airplane lies upon it. To give the airplane a climbing effect, tilt the bottom support upward as shown in Fig. 7-1.

A simple support rod may be constructed from a 14 inch piece of 1/16 inch brass tubing (Fig. 7-8). Form a loop in one end of the brass rods so that the weight will be balanced under the wing area. Bend the two ends downward so they may be soldered to the top of the tower. The plane should ride about five inches above the tower. First, bend the loop so the plane will have a slight climbing incline at the front end. Solder the support rod to the brass tower. Straighten and level the model support rods. Now, apply a thin layer of clear silicone cement on top of the support rods. Place the plane in position. Circle two small rubber bands around the airplane body to hold it to the support rods until the cement has set up. Pull the motor wire down through a small hole in the bottom of the plane area. Use the sharp point of a soldering iron to make a new hole for the motor leads. Press the wires on the support rods and solder one lead to the tower. Solder the red lead of the motor to the small wire pulled up through the tower leg. Taping up the exposed connections is not necessary. Apply a light coat of silicone rubber cement over the connection. Simply place the airplane in the sun and watch it fly.

PROJECT 8
SOLAR IC AMP

A simple audio amplifier with a minimum of parts may be constructed around the eight prong LM 386 IC circuit. You may amplify a crystal set or transistor broadcast receiver with this amplifier. Although the speaker may not blast your ears, the volume is quite adequate. This little amplifier may be built on an experimenter board or perfboard (Fig. 8-1).

Since the LM 386 will perform with low voltage and current drain, the amplifier is ideal for either battery or solar operation. Here, a 6 volt supply voltage from fifteen solar cells works nicely. The current drain is normally only 5 mA and with a strong signal goes up to only 11 mA. All solar cells are connected in series to obtain a 6 volt supply. Only seven components are used in this small solar amp. See Table 8-1.

Construction

For experimental purposes the solar IC amp may be constructed on an experimenter board. The amplifier can be put to-

Fig. 8-1. This little solar amp may be constructed on an experimenter board in 30 minutes. Just plug the components into the board area.

gether in thirty minutes or a little longer. If the amplifier is to be used constantly, it should be built on a perfboard. An eight-prong IC socket is glued to the board after all components are soldered in place.

Select a 3 × 4 inch perfboard or cut your own from a larger piece of stock material (Fig. 8-2). Make an L-shaped volume control bracket out of a piece of scrap metal. Drill a ⅜ inch mounting hole for the volume control and two small ⅛ inch chassis holes. Mount the bracket and volume control to the front of the perfboard. The

Table 8-1. Parts List for Project 8.

```
IC — LM 386 — Radio Shack #276-1731
Perfboard — 3×4 inches cut from larger stock —
              Radio-Shack #276-1394,
              GC Electronics #J4-602
Foam Chassis — 6 × 9 inches of 1 inch foam material
SPK — 8r 3 to 4 inch PM size
C1 — .05 250 volt fixed capacitor
C2 — 470 of 16 V electrolytic capacitor
C3 — 220 of 16V electrolytic capacitor
R1 — 10K miniature volume control, GC Electronics #G1-662 —
              ETCO #068VC
Solar cells — 15 total — 100 mA — Solar Amp #S122,
              Poly-paks, Inc #5306, H & R Inc. TM 21k666,
              John Meshna, Inc. H-14A
Misc. IC 8 pin socket, hookup wire, solder, mounting bracket
              silicone cement, etc.
Cost — $35.00 to $37.50
Time — 6 hrs.
```

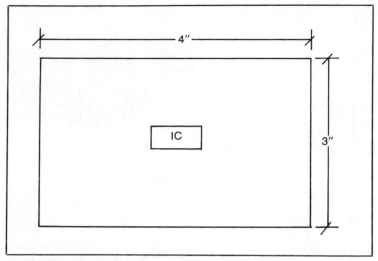

Fig. 8-2. Cut a piece of perfboard 3×4 inches from a larger piece of stock. This small perfboard chassis will hold all components except for the speaker and solar cells.

small speaker may be held in position with a small section of foam and silicone cement.

First, mount all electronic components on the perfboard. Small components may be mounted as they are wired into the circuit. Eventually, the speaker and perfboard will be mounted on a piece of 1 inch foam chassis. Mount the solar cells at the rear of the foam chassis. The speaker and perfboard may be mounted on the foam chassis after the solar cells are mounted and tested.

Wiring the Circuit

Start with the volume control and solder the two leads. Run a bare piece of hookup wire along the bottom edge. Hook each end into the small holes of the perfboard. This bare wire will serve as a common ground wire. Twist the two volume control wires together and connect the ground end to the bare ground wire. Solder the center terminal wire of the volume control to pin 2 of the IC socket (Fig. 8-3).

Solder a wire to pins 3 and 4 of the IC socket to the ground wire. Connect an 8 inch piece of hookup wire to pin 6 to go to the positive (+) terminal of the solar supply. Mount a 470 μF capacitor from pin 6 to pins 3 and 4 of the IC socket. Try to mount all components as close to the IC socket as possible.

Solder C3 (220 μF) to pin 5 of the IC socket. Run both capacitor leads through the perfboard and back up so these larger capacitors will not break off. Connect two ten inch pieces of hookup wire for the speaker. Solder one lead to C3 and the other to the common ground wire. Twist these two wires for a neat appearance (Fig. 8-4).

After all components have been soldered to the IC socket, use rubber silicone cement and bond to the perfboard. Connect a .05 μF capacitor to the high end of the volume control (R1). The input signal will connect at this point. C4 acts as a dc blocking and signal coupling capacitor.

Mounting the Solar Cells

The fifteen solar cells are mounted to the rear of the foam chassis. Keep the final two wire connections close to the perfboard chassis. Although the small amplifier pulls very little current, 100 mA or larger cells were chosen for operation. The larger cells collect more light and prevent motorboating within the amplifier circuits.

Solder a ⅛ inch piece of flexible wire to the bottom of each solar cell. Cement the cells to the foam chassis with silicone rubber cement. Use either clear or white rubber cement. These colors do not smear up the white foam area. Press the cells down onto the foam. Keep the cells in a straight and neat fashion. Let the cement set up overnight.

Connect each wire to the next cell at the center tie bar. All cells are wired in series (Fig. 8-5). Connect a six inch piece of wire to cell number 1 and cell number 15. The positive lead from cell number 15

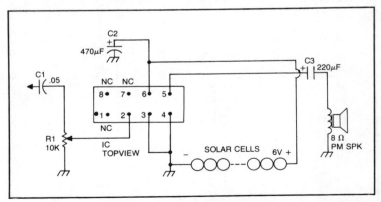

Fig. 8-3. Only seven components are found in this small solar amp. The amp pulls 5 to 11 mA of current under working conditions.

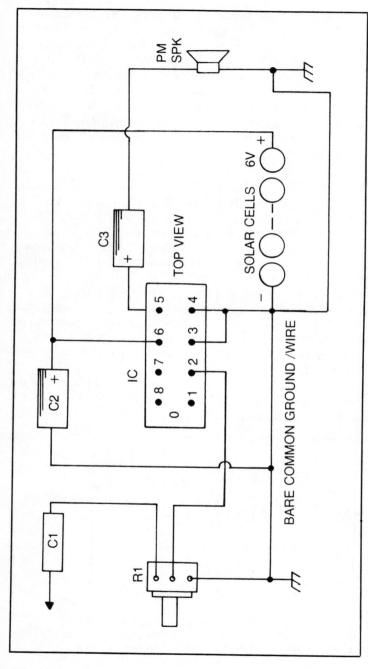

Fig. 8-4. This is a simple wiring diagram of all the components for the IC solar amp. Keep all leads as short as possible.

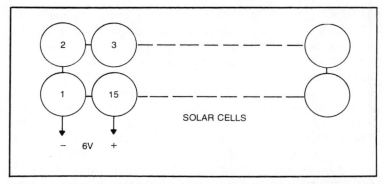

Fig. 8-5. Connect all cells in series. The solar cells are mounted on the rear of the foam chassis.

will be on the bottom side of the cell. Check the solar cell voltage before mounting the perfboard and speaker.

Under a 100 watt bulb you should measure between 4 and 5 volts. Setting in the sun, the voltage should be over 7 volts. Suspect poor soldering contacts or a defective cell if you have lower voltages. Double check the wiring. Locate any poor connections or cells with a voltage measurement across two cells. The poor connection or defective cell will have a lower than normal voltage reading. Inspect each cell for cracks or broken sections.

Mounting Large Components

The perfboard amp should be mounted close to the front of the foam chassis. Make all connections before cementing in place. Cut and solder each solar cell connection to the perfboard. Make sure the top side connection of cell number 1 goes to the common ground connection. The positive wire should go to pin 6 of the IC. Now, connect both speaker wires. The speaker will be mounted to the right of the perfboard.

Cement the perfboard to the foam chassis. Use silicone rubber cement in each corner of the amplifier board. Line up the perfboard chassis and press in place. The small pm speaker may be mounted in the same manner. Place a piece of foam material under the speaker magnet. Cement both components to each other and the foam chassis.

Testing

After the rubber cement has dried, the small IC amp may be tested. Place a 100 watt bulb over the solar cells. Grab hold of the end of C1 and you should hear a loud hum in the speaker. If the volume control is set too high, you may hear motorboating or a put-put noise in the speaker. When placed in the sun all motorboating ceases. Usually, motorboating occurs when not enough light strikes the solar cells.

Take a voltage measurement at the output of the solar cells when no hum is heard. The voltage reading should be from 4 to 6.5 volts. Very low voltage may indicate improper wiring of the IC. Place a mA meter in series with the positive lead from the solar cell and amp connection. Set the vom range at 30 mA. With no signal, you should measure between 3 and 5 mA. Now, touch the end of C1. The current should jump to around 11 or 12 mA. A loud hum in the speaker indicates the amp is working.

You can now connect that crystal set or transistor radio to the solar IC amp. This little amp may be used as a radio signal tracer with radio or cassette players. Simply connect a ground lead to the chassis of the unit to be tested. Connect another lead to C1 and start at the volume control of the defective unit (Fig. 8-6). The audio signal may be traced through the audio stages until the signal is lost. The defective component is then close at hand.

Mount all parts as close to the IC socket as possible. Use short leads on the input connections. Twist the two input leads and mount close to the IC. The two solar power leads should be twisted to make a neat appearance. Mount all solar cells close together.

Fig. 8-6. Here the solar amp is checking the audio signal of the small battery powered AM radio. You may also use the signal from the radio to test the amplifier for volume distortion.

The automatic night light may be used in the children's or baby's bedroom. When the sun comes up, the night light is automatically switched off. Likewise, with no outside light, the unit switches on the inside light. This automatic solar night light may be used to turn on a reading lamp in the living room (Fig. 9-1). When you are on vacation it can prevent someone from burglarizing your home. The automatic device should be placed where the outside light will strike the topside solar cell.

Only 6 Vdc is needed to operate the small relay. The relay may be operated from batteries or a rectified ac source. Here a 6 volt ac power transformer was used with a silicon diode and filter to provide a dc source to the relay. A 1 volt solar cell is wired in series with the solenoid relay winding. Internal resistance of the cell is almost 100% when not exposed to light so no voltage is applied to the solenoid. When light strikes the solar cell, the internal resistance drops permitting the 6 volts to reach the relay which energizes and turns off the inside light (Fig. 9-2).

Use a 100 watt bult or less when using the solar night light to turn on the lamp in the living room. A 60 watt bulb is ideal. Select a 7½-or 10 watt bulb for the night light. See Table 9-1.

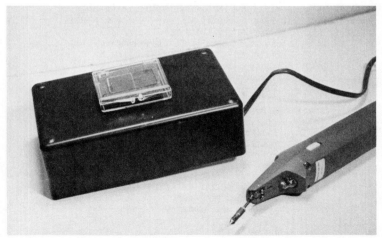

Fig. 9-1. The automatic night light may be used to turn on a light in the children's bedroom or to turn on a reading lamp in the living room.

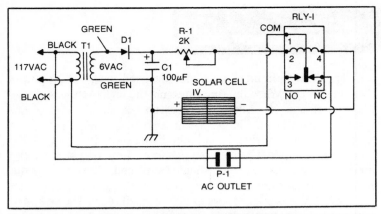

Fig. 9-2. A low voltage power supply is used to energize the solenoid winding. The solar cell controls the voltage applied to the relay.

Construction

Mount the solar cell in the center of the plastic mini-box. The solar cell may be left right in the protective plastic box it comes in. Simply drill two small holes in the bottom of the plastic box. These same holes should be drilled through the mini-box. Place the cell inside the plastic box and pull the two leads through the mini-box.

Table 9-1. Parts List for Project 9.

```
T1  — 117 V ac to 6 volt filament power transformer
        GC Electronics #D1-745, Radio-Shack #273-1380
        GMI #D1-745, ETCO # 037xF
D1  — 1 amp silicon diode — GC Electronics #J4-1600
        Radio-Shack #276-1102, ETCO #007RC
R1  — 2K ohm control, Screwdriver adjustment
        GC Electronics #B1-643, ETCO #318VA
RLY — 1 Relay Radio-Shack #275-004
        (6-9V relay with 1 amp 125 Vac contacts)
Solar Cell — 1 volt or two .5 volt cells wired in series.
        GC Electronics #J4-801 (1 volt solar cell).
        3 — .5 volt cells, Poly-Pak #5306
        GC Electronics #24-800, Edmund Scientific Co. #42,268
        Solar amp #S122, H & R, Inc. #TM 20K187
        John Meshna, Inc. #H-12
Mini-box 5×2½×1⅝ — GC Electronics #H4-722
        Radio-Shack #270-231, GMI #5LTF778
        ETCO #159VA or equivalent
Misc. — Ac plug, ac wire, solder, hookup wire, etc.

Construction time — 4 to 6 hrs.
Cost — Under $15.00
```

Cement the bottom of the plastic box to the mini-box with silicone rubber cement.

The controllable light may be wired directly into one end of the plastic mini-box or install a female ac plug. Drill a ¼ inch hole for the ac cord in the opposite end of the plastic box. The small power transformer should be mounted in the backside of the box with nuts and bolts. Mount the relay with a dab of silicone cement on the bottom side and stick it to the mini-box.

Since only bare wires extend from the small relay, connect hookup wire with insulation over the bare connections. Check the bottom wiring schematic for correct terminal connections (Fig. 9-3). The solenoid winding terminals are in the center. Connect the common terminal to one side of the ac line and the NC terminal to the ac plug or light.

The Power Supply

A half-wave rectifier circuit is established with a 6 volt step-down transformer, silicon diode, and filter capacitor. The voltage output is approximately 10 volts. R1 is adjusted between 7 to 8 volts. All power supply components are mounted on the plastic mini-box, instead of the metal plate. This prevents any type of ac grounding of the power line and possible shock hazard.

R1 is a small wire-wound adjustable bias control. Although this small control may be located at most Radio & TV wholesale stores, any type of resistance control may be used from 2k to 5k. Correct adjustment of this control will determine the sensitivity of the solar cell and relay. R1 should be mounted on the plastic area with a small ¼ inch hole in the center for a screwdriver adjustment.

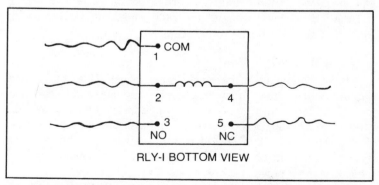

Fig. 9-3. This little relay comes with bare connecting wires. Here is the bottom view of the connection terminals.

Wiring the Circuit

After all large components are mounted, start connecting the ac circuits. Insert the ac cord in the ¼ inch hole of the plastic mini-box, and tie a knot about 8 inches from the end of the cord. This will prevent the cord from being pulled out of the box. Solder both black leads of the transformer to the ac line cord. Connect a wire from one side of the power line to the ac light plug. Now, run another wire from the other side of the power cord to the (com) terminal of the relay. Slip a piece of insulation over this connection. Tape up all ac wiring connections.

The secondary winding of the power transformer (6V) goes to common ground and the silicon diode. Use a four lug terminal strip to hold the diode and small electrolytic capacitor (Fig. 9-4). Connect the positive terminal of the diode to the positive terminal of the capacitor (C1). Run a piece of hookup wire from this connection to one side of R1.

Next, connect all wires to the small relay. Although, the relay terminals are only lug wires, they were numbered on the schematic for easy reference. Connect a piece of hookup wire from R1 to terminal 2 of the solenoid winding. The black wire (−) of the solar cell was soldered to terminal 4 of the solenoid. The red wire (+) from the cell is soldered to the common ground terminal. If the solar cell has no connecting wires, solder a red lead to the bottom side of the cell (+) and a black wire to the top side (−). From the normal closed (NC) terminal of solenoid (5), connect a wire to the existing side of the ac plug.

Double check all wiring connections before the unit is plugged in. Make sure all ac connections are taped or covered. Place rubber silicone cement (after connections are made) on the female ac plug (P1). All bare connections of the relay should be covered with insulation and/or silicone rubber cement.

Checking the Unit

Before plugging in the power cord, adjust R1 to the center rotation of the control. If a vom is handy, clip one lead to common ground and R1. Insert the power cord and adjust R1 for 7.5 to 8 Vdc at terminal 2 of the relay. No voltage may indicate improper wiring of R1, D1, or C1. You should measure 10 Vdc at the positive terminal of the silicon diode. Check over the component wiring in the low voltage power supply.

Turn the plastic box over and set the unit in the sunlight. Plug a reading lamp or night light into the female ac jack. The light should

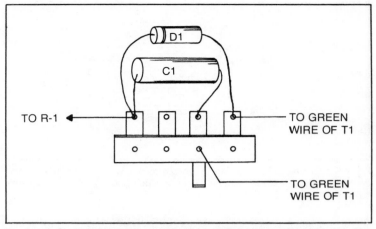

Fig. 9-4. Use a four lug terminal strip to mount the diode and filter capacitor. The green secondary wires of T1 are soldered to the terminal lugs. If the transformer has a center-tip lead, cut it off and tape it up.

be off. Now cover up the solar cell with your hand and the light will come on. When the solenoid seems sluggish to turn on the light, adjust R1 for higher dc voltage. If R1 is turned too high, the light may stay on. Adjust R1 until the light goes off and on when light strikes the cell or the cell is covered up.

When In Trouble

An inoperative solar night light may be caused by improper wiring of the low voltage power supply. Check the voltage at the positive terminal of the silicon diode (10 Vdc). If okay here, go to the solenoid. You should have 7 or 8 volts at pin 2 of the solenoid winding. No voltage here may indicate R1 is open. With a 100 watt light over the solar cell, the solenoid should energize. Measure the voltage across the solenoid winding. This solenoid will start to operate at 6 volts.

In case the silicon diode is installed backwards, the circuit is inoperative. Feel the small capacitor for warmness. Improper connections of the diode may destroy the capacitor, diode, or power transformer. A reading of 50 to 150 ohms across the ac plug (out of the power socket) indicates the primary winding is normal. Often when a direct short is found across the secondary circuitry, the primary winding opens also.

The night light will come on when the sunlight hits the solar cell with improper terminal connections at the relay (NC). The

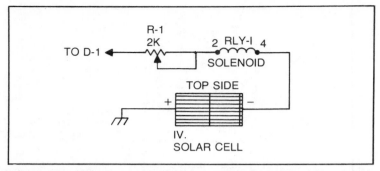

Fig. 9-5. When the solar cell is wired backwards, the night light will never come on. Make certain that the negative lead (top side of cell) goes to the solenoid winding.

normal open (NO) terminals are not used in this project. Terminal 5 or the normal closed (NC) terminal should be connected to the ac outlet or night light.

What happens when the solar cell is wired up backwards? (Fig. 9-5). The night light will never come on. Since the cell is wired backwards, the solenoid becomes energized without any bucking voltage from the solar cell. Check the solenoid and notice if the reed switch is pulled downward. Simply reverse the leads to the solar cell.

PROJECT 10
FLASHING SOLAR CONTINUITY CHECKER

This little solar continuity checker will test house fuses, iron or toaster elements, motors, light bulbs, or just about any electrical or electronic device you have around the house. If the resistance of the product or component you are checking is under 100 ohms, the LED will flash off and on indicating continuity of the circuit. You may use the flashing continuity checker for locating broken or open wires.

The flashing solar continuity checker is constructed around a flashing LED. Actually, the component looks like any red LED except it has a built-in flasher IC. The flashing LED is powered with several solar cells and has test leads (Fig. 10-1). A typical 5 V supply voltage at 20 mA of current will operate the flashing LED. When higher voltage than 5.5 volts is applied, the bulb may stay on or flash slower than usual. The LED has a typical pulse rate at 5 V of

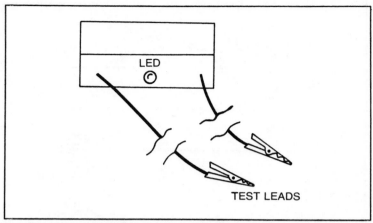

Fig. 10-1. The small LED continuity tester may be constructed in a clear plastic or mini-box. Keep the LED down inside and out of the lighted area.

3Hz. As the tester is held closer to the light or under bright sunlight, the pulse rate of flashing slows down. There are no batteries to wear out and no shock hazards to the operator when operating this continuity checker. See Table 10-1.

With only 5 to 6 volts and 20 mA of current, any small solar cells will fill the bill. Choose fourteen solar cells, with a current rating above 25 mA (Fig. 10-2). These solar cells may be broken pieces, chips, or crescent type cells. If you have a few broken ones, connect these cells in series to get the required operating voltage.

Choosing the Container

The small continuity checker may be constructed inside a plastic box, mini-box, or on a piece of wood as Project 6. Just make sure the solar cells are protected since the continuity tester will be used a lot around the house or shop and should be able to stand a few

Table 10-1. Parts List for Project 10.

```
14  —  small solar cells, 25 mA's of current or greater. May be chips, pieces or crescent type cells.
       Crescent solar cells — #P-42,749 — Edmund Scientific Co.
       One quarter solar cells #S122 — Solar Amp, Inc.
       Broken cell and pieces — #H-13 — John J. Meshna, Inc.
       Square solar cells — #TM20K187 — H & R, Inc.
1   —  Plastic Box 4×3×2 or equivalent
2   —  Small test clips
1   —  Flasher LED — #276-036 — Radio Shack
Misc.— test lead wire, small cell connecting wire, piece of foam
       material and clear silicone cement

Construction time — 4 hrs.
Cost of project — less than $12.00 to $25.00.
```

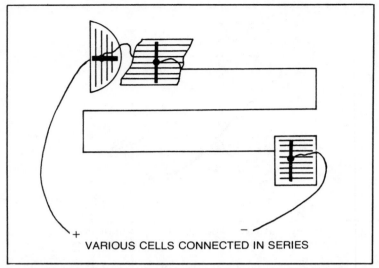

VARIOUS CELLS CONNECTED IN SERIES

Fig. 10-2. Connect fourteen solar crescent pieces or chips in series for the required 5 to 6 volts. Practically any size cells will work.

knocks. If the solar cells are placed inside a plastic box, mount the cells on a piece of foam material (Fig. 10-3). This provides insulation, easy cutting, and substantial mounting area for the solar cells.

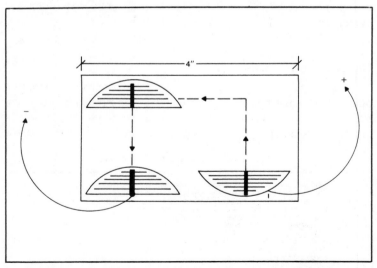

Fig. 10-3. Mount the cells on a thick piece of styrofoam. Cut the foam to fit snugly inside the clear plastic box.

If the solar cells are to be mounted inside a plastic mini-box, place the solar cells inside and mount them on a piece of clear plastic. With this method, you must apply clear silicone cement to the front of the cells to hold them in place. Don't forget that the negative terminal wire must be soldered to the front of each cell before mounting.

When using a clear plastic box as a container, cut a piece of foam so it will fit inside the plastic box. This way the solar cells will be protected and the two continuity test leads may come directly out of the front of the box. The small LED may mount at the bottom edge between the two test leads so the light will be protected from the strong sunlight or desk lamp. Actually, you can see the light flash a lot better.

Mounting The Solar Cells

Crescent solar cells work nicely for this project. They may be mounted lengthwise or in two separate rows and then connected together. The plastic box should be no smaller than 2 inches × 4 inches. Cut the piece of foam so the lid of the box will fit tightly down on the solar cells. Now, the crescent cells may be mounted on the foam area.

First, solder a small wire to the middle bottom side of each solar cell (Fig. 10-4). Choose very small connecting wire or take a strand of copper wire from an antenna cable or ac cord. A small strand of wire, 1 inch long, is sufficient. Scrape the back side of the cell and solder a 1 inch piece of wire to each cell. Bring the wire out and bend it upwards. After cementing into position, these loose wires will solder to the next cell.

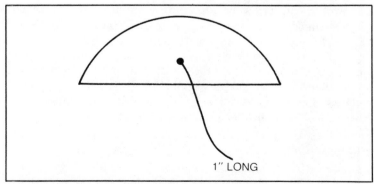

Fig. 10-4. Solder a small wire to the middle area of the bottom side of the cell. This side is positive (+) and has a shiny silver area.

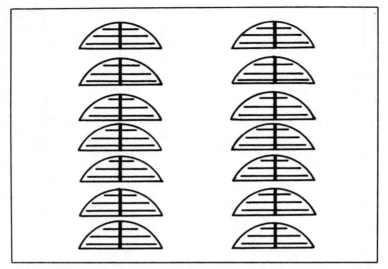

Fig. 10-5. Keep the cells ⅛ inch away from each other. Run them in two rows.

Now, lay out the cells on the piece of foam material (Fig. 10-5). The cells may be only ⅛ inch away from each other. If the plastic box is only 4 inches long, make two rows of cells. With this layout, make sure the 7th cell has a 2¼ inch length of wire to connect to the next row. The first row of cells should be mounted in a row with the round part at the top and the second row upside down to connect all cells in series. If you desire to have all cells going the same way, the connecting wire of the 7th cell must be about 6 inches long to reach the top section of cell number 8 (Fig. 10-6).

After a connecting lead is soldered to the bottom side of each cell, with a correct layout, place a bead of clear silicone cement down the middle of the two groups of cells. This thin line of cement will hold the cells into position. Only a small bead of cement is necessary. Too much cement will ooze up around the solar cells. Now, push each cell down on the cement. Keep the cells in line and square them up. Place the cells within ⅛ inch of each other so all the cells will fit as planned. Make sure the small connecting wire is clear above each cell. Let the cement set up overnight.

Wiring the Unit

Start with the top cell and extend the wire lead down through the piece of foam. The wire connection doesn't have to be taped since the connection will be inside the foam area. Start with the first

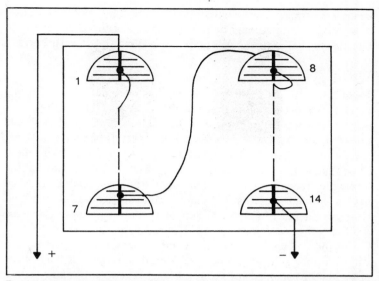

Fig. 10-6. If you desire to keep the crescent cells uniform (all in one direction) the connecting wire of the 7th cell must be about 6 inches long to reach the top section of cell number 8.

cell and connect to the second, etc. Solder each piece of wire to the center terminal (top) of each cell until all cells are wired in series. Now, connect a four inch length of wire to the top of the last cell. Run this wire down through the foam and connect to the anode terminal of the LED (Fig. 10-7).

Connect a ten inch test lead and clip to the cathode terminal of the LED. Tie the wire into a knot before feeding through the plastic

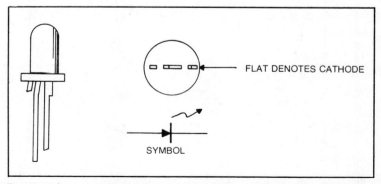

FLAT DENOTES CATHODE

SYMBOL

Fig. 10-7. Connect the topside (−) of the last solar cell to the anode terminal of the flashing LED. Here is the bottom view of the LED with the correct symbol.

cabinet, so the test cable will not pull out. Solder another test lead and clip to the remaining negative terminal of the first solar cell. Double check the wiring. All components should be wired in series.

Testing

Check the tester after all wiring has been completed. Simply clip the two test clips together and either set the tester in the sun or under a reading lamp. Look straight at the LED. The LED should blink off and on. If not, reverse the leads to the LED.

When greater problems arise, take a voltage measurement across the entire solar pack (Fig. 10-8). Measure the voltage at the lead which feeds the LED and the positive clip test lead. You should have a voltage over 5 volts under a reading light. Hold the light closer to the unit. As the light gets closer, the voltage will become greater.

If the voltage is less than 3 volts, check each cell connection. Either a poor solar cell, poorly soldered cell, or poor connection prevents the required voltage. Check across two cells and compare this voltage with the next two. Double check the polarity of the LED.

Once the LED starts to flash, move the light closer to the solar cells. When the light is too close, the LED flashing slows. Now, place it very close. The LED may stay bright and not flash. With

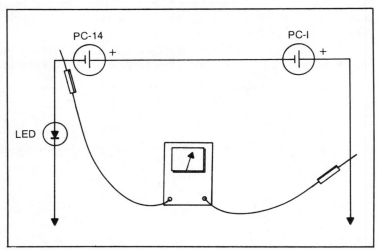

Fig. 10-8. When trouble arises and the LED will not light, connect a voltmeter across the entire solar pack. Measure the voltage before the anode terminal of the LED.

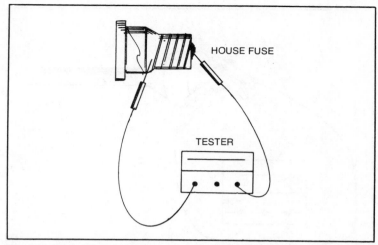

HOUSE FUSE

TESTER

Fig. 10-9. Clip or touch the two metal areas of the fuse with the small test clips. If the fuse is good, the LED should flash on and off. No flashing indicates an open fuse.

your hand, cover up part of the solar cells and notice the flashing of the LED. As more cells are covered, the LED becomes dimmer and dimmer.

Don't overlook light hitting the LED to prevent it from lighting up. The sunlight or overhead light when striking directly upon the LED may keep the unit from flashing. The LED was placed low to give out more light to the operator. Enclose the top and sides of the LED to prevent light from striking the body and producing greater flashing. Use a piece of black spaghetti insulation or silicone rubber cement over the outside body of the LED. The same rubber cement may be used to seal and hold the LED in position.

Connect a small voltmeter across the LED. As the LED begins to flash, the hands of the voltmeter will move up and down on the voltage scale. Notice the change of the meter hand as the light is less, or as part of the solar cells are covered up.

Now, the flashing continuity tester is ready to check our a few items around the house. Low resistance items, such as fuses, cords and wire connections may be checked by clipping or touching the two lead wires across the component (Fig. 10-9). Simply hold the clips across the suspected house fuse. What happens when you are in the basement trying to locate the defective fuse without any sunlight and it is pitch dark? Hold the flashlight down against the top of the solar cells and the little tester flashes on. A three cell

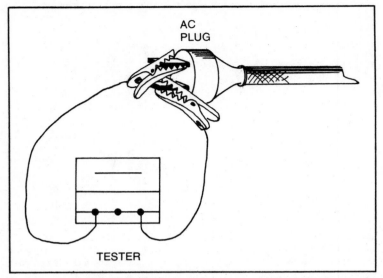

Fig. 10-10. To check a defective toaster or appliance, clip the test lead across the power plug. *All continuity tests are made with the power cord removed from the wall outlet.*

flashlight or any wide flashlight lens makes the LED flash off and on for continuity tests.

To check out a defective toaster, clip the two leads across the ac plug (Fig. 10-10). When making continuity tests on any appliance, *make sure the ac cord is pulled out of the wall outlet.* Now turn the toaster on. The light should blink if the cord, switch and heating elements are normal. No light will indicate a broken cord or element. If one side of the toaster works and the other does not, dismantle the toaster and check the continuity of the side which does not light up. You may find a broken wire element.

To test small motors, such as a knife sharpener, mixer, or slicer, clip the two leads across the ac plug. Push down the switch and the flashing light should come on. If not, either the motor field winding is open or suspect a broken switch and cord. Most likely, you will find a broken cable between the ac plug and appliance. Dismantle the appliance or take off the bottom panel to get at the inside ac connections.

To check out the ac cord test one side of the cord at a time. Clip one lead to the one side of the ac plug and touch the other clip to the end wire on the cable (Fig. 10-11). Touch both wires for this test. When the light begins to blink, the wire continuity is normal. Now,

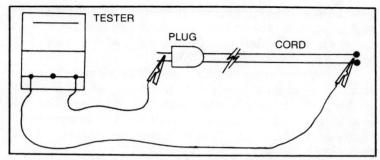

Fig. 10-11. To check a power cord, clip one lead to the ac plug tip and to the corresponding wire within the appliance. Continuity is good when the light flashes off and on. No light indicates a broken wire within the power cord.

Fig. 10-12. This is the schematic for the continuity tester.

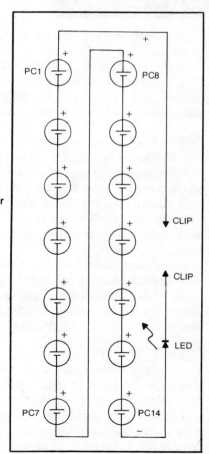

go to the other side. Place the test clips on the opposite prong of the ac plug and touch the remaining wire. If the light does not come on, the wire is broken. Most ac cords break at the ac plug or where the cord enters the body of the appliance.

This little flashing continuity tester may prove to be a very practical test instrument around the house. You can just about test anything in the home for continuity tests. It's all done with a little outside light—no batteries to wear out. It works every time and will last for years. The wiring schematic of the little continuity tester is shown in Fig. 10-12.

PROJECT 11
SOLAR WINDOW RADIO

Just snap this solar radio to the window and listen to your local broadcast station. There are no batteries, switches, or tubes to wear out. This little radio is powered by three solar cells. On strong local stations, no outside antenna is needed. For more difficult stations, add a 12 ft. piece of flexible antenna wire. Very difficult stations may be received with an outside antenna.

How It Works

The broadcast signal is picked up by a long antenna coil (L1) and the stations are tuned in with C1 (Fig. 11-1). Transistor (Q1) acts as a regenerative detector with feedback coil L2. L3 is rotated until a loud squeal is heard on a local station. Separate the stations with C1. Back the regeneration control (R3) off until the squeal quits. Now, you can hear a station in the small earphone. See Table 11-1.

The audio signal is transformer-coupled to capacitor C5. C5 serves as a dc blocking and coupling capacitor. The signal is further amplified by transistor Q2. Three small solar cells, wired in series, furnish the dc voltage to the radio circuits. You may let the radio play all the time. No on/off switch is needed here.

Preparing the Board

The radio is built on a 4½ × 6 inch piece of perfboard. These boards can be found at most local radio supply houses. In each corner drill a ⅛ inch hole to mount the suction cups (Fig. 11-2). Drill a ⅜ inch hole to mount the variable capacitor and regeneration control. All other components are mounted through the small perforated holes. Keep all components on the window side.

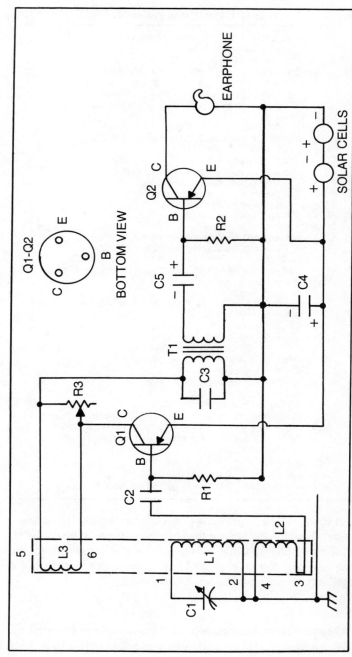

Fig. 11-1. This is the schematic of the radio that receives its power from the sun. A regenerator type circuit provides local and some distant reception of broadcast stations.

93

Table 11-1. Parts List for Project 11.

```
L1 — Ferrite antenna coil to match 365 pF variolle capacitor
     or wind your own.
L1 — Winding your own
       1 — MS081 Ferrite rod (8″) ETC Electronic Corp.
             L1 — 72 turns over ⅜ inch ferrite rod #26
             L2 — 26 turns ⅛ inch from L1 #36 enamel wire
             L3 — 10 to 12 turns #36 enamel wire
C1 — 365 pF variable capacitor
       GC Electronics A1-233
       ET Co Electronics Corp. 185VA or 042cc
C2 — 101 pF capacitor 50V
C3 — .005 pF capacitor 50V
C4 — 100 mfd 12 volts electrolytic capacitors
C5 — 1 mfd 12 volt electrolytic capacitor
R1 — 1.2 megohm ½ watt carbon resistor
R2 — 100K ½ watt carbon resistor
R3 — 50K ohm variable control
Solar Cells — 3-50 mA solar cells
       Solar amp #S122, Edmund Scientific Co. #30,735
       Poly Paks, Inc. #5306, Radio Shack #276-125
       John Meshna, Inc. #H-11, H & R Inc. #TM 20K187
       GC Electronics J4-801 (2 only required — 14 solar cells).
T1 — Driver Transformer
       Argonne TR97
       GC Electronics DL-728 1000rpn — Sec 500 r impedance
       Gravois Merchandiser, Inc. — 13A833.9
       ET Co Electronic Corp. — #V02xF or equivalent
Q1 — Q2 — 2SA52
       RCA SK-3004, GE — 2, Radio Shack
       276 - 2007 or 276-1604 pnp Germanium,
       GC Electronics 24-1625
P1 — 1000-2000 ohm impedance earphone
       GC Electronics #J4-825
       ET Co Electronics, Inc. #272VA
J1 — Miniature earphone jack (fits earphone above)
Suction Cups — 4 1½ — picked up at local hardware stores
1  — Perforated board 6×4½ inches — GC Electronics #J4-602,
       Radio Shack 276-1394
Misc. — hookup wire, solder, knobs, etc.
Construction time — 12 to 14 hours
Cost — under $25.00 (with some parts from the scrap box).
```

First, mount the larger components. Wrap hookup wire around the two ends of L1 and secure through the perforated holes. Next, mount the variable capacitor and volume control. Do not mount the suction cups until the radio is functioning. The three solar cells are mounted after all wiring has been completed.

The Antenna Coil

A regular ferrite antenna coil may be used that will tune with a 365 pF capacitor (C1). One with a longer coil is better to pick up

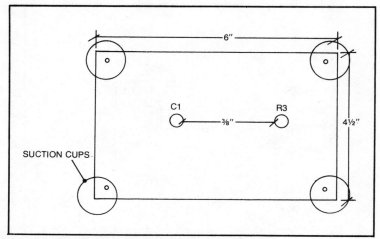

Fig. 11-2. Build the small window radio on a perfboard.

local broadcast stations. If L1 is not handy, you can wind your own. Pick up 7 ft of number 26 and 5 ft of number 36 enameled wire. This wire may be taken from other coils or old transformers. When using a commercial coil, just wind L2 and L3 over the original windings.

If starting from scratch, select a 6 to 8 inch ferrite rod and wind the coils on it. First, wind L1 close wound on the ferrite core (Fig. 11-3). Now, over each end of the coil wind L2 and L3. Place a layer of scotch tape between the coil layers. Scotch tape may be added over the windings to hold the wire in place. Scrape the enamel coating off of each wire end and tin the scraped area. A very poor soldered connection will result if the enamel is not scraped away. Now, the antenna coil can be fastened to the perfboard.

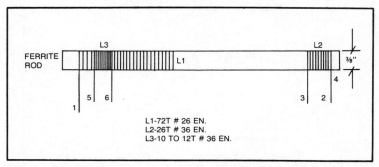

Fig. 11-3. How to wind your own broadcast coil. If a commercial coil is not handy, you can wind your own.

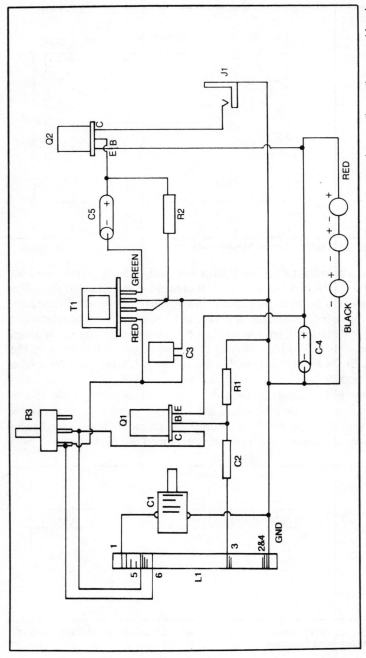

Fig. 11-4. If you have problems wiring the small radio, follow the pictorial layout. Cross off each component and connection as they are soldered.

Wiring the Radio

After all large components are in place, the radio may be wired starting with the antenna coil. Connect terminal 1 of L1 to the station plates of C1. Solder terminal 2 to the ground or rotor terminal of C1. Connect all of the ground connections to the ground rotor terminal of C1. Connect one end of C2 to terminal 3 of L3 and the other end to the base terminal of Q1. Terminal 4 of L3 solders to the ground connection of the variable capacitor. Following the pictorial diagram (Fig. 11-4) to connect T1 and Q2.

A phone jack may be used to disconnect the headphone or it may be wired permanently into the perfboard. Tie a knot in the earphone cord when soldered directly into the circuit. T1 may be held to the perfboard by bare hookup wire soldered to the front of the transformer. The small components may be mounted as they are wired in the circuit.

Connecting the Solar Cells

Before mounting and wiring the three solar cells, double check the radio wiring. Go over the circuits at least twice and mark off each component as it is connected into the circuit. Although the voltage from the solar cells will not destroy the transistor, it's very discouraging to wire up a project and it doesn't work the first time.

You may connect the three solar cells in series before cementing to the perfboard (Fig. 11-5). If the cells have connecting wires, take the red wire and solder to the black wire of the next cell. The top wire of the solar cell is always negative. Tie the three cells in series for the 1.5 volts. Greater volume may be possible with four solar cells in series.

Single cells without wires may be added in series with a wire soldered to the bottom side (+) and then connected to the topside of the next cell. A three inch lead should be connected to the topside of

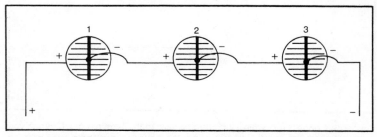

Fig. 11-5. Connect the three solar cells in series. Solder the bottom side wire to the topside of the next cell.

number 1, and the bottom side of the number 3 cell. After wired in series, place a dab of silicone rubber cement on the backside. Lightly press the cells on the perfboard. Wipe off any excess cement that may come through the holes of the perfboard.

Now, solder the negative terminal (top wire) to the common ground of the variable capacitor. The remaining cell wire (+) should be soldered to the emitter terminals of both transistors. The cells are connected into the circuit at all times. There are no switches to turn the radio on and off and no batteries to wear out.

Check Out

When all the wiring and solar cells are completed, the radio may be tested. Plug in the earphone. Hold the radio up to a window or under a reading lamp. Lay the small antenna wire around the room. Now, turn the regenerator control up until you hear a squeal. If no squeal, reverse the two coil leads of L3.

Rotate the control until you hear a squeal in the earphones. Turn the variable capacitor to a local broadcast station. Slowly back the regenerator control down until the whistling ceases and the station plops in. Readjustment of the control and tuning capacitor may be needed for good reception. Slowly tune in other stations.

The radio can now be completed by mounting the four suction cups. Simply add a drop of water or wet each cup and stick it to an outside window. You will find that with a long antenna coil, several

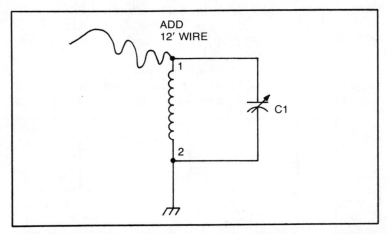

Fig. 11-6. Connect a flexible 12 ft antenna wire to the terminal of L1. When you live several miles from a broadcast station the added wire will help bring in these stations.

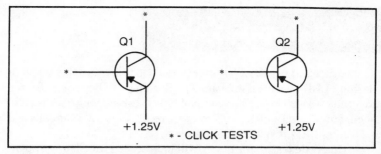

Fig. 11-7. Here are some click and voltage test points to check when the radio will not perform. Don't forget to reverse the two coil leads of L3 when the radio will not operate.

local stations may be heard. The antenna will help pull in distant stations (Fig. 11-6).

When the receiver seems dead or will not regenerate, take several voltage and click tests (Fig. 11-7). Place the radio under a 100 watt bulb and plug in the earphone. Take a small screwdriver blade and touch the base terminal of Q2. Be careful not to short out any of the transistor terminals. You should hear a click or hum if Q2 is amplifying. If not, check the wiring and terminal connections of Q2. It's possible too much heat may have been applied to the transistor leads (destroying Q2) when soldering.

If a click or hum is heard, proceed to the other end of C5. Now, go to the blue lead or collector terminal of Q1. The click should be heard. Touch the base of Q1 and the hum should be louder. No sound may indicate a poor connection or defective transistor. Check for a positive voltage on the emitter terminal. This voltage is very low, somewhere between 1.25 and 1.45 volts.

After voltage and click tests are made, you should be able to locate the defective area or component. These small transistors may be ruined if too much heat is applied to the terminal. Leave the terminals long and use a small low-watt soldering iron. For added protection, use a pair of long nose pliers or a heat sink. Grasp the terminal that is to be soldered with the long nose pliers at the transistor body. The pair of pliers will absorb any heat applied to the transistor terminal.

Although this radio will not hurt your ears with volume or pick up as many stations as a commercial unit, it's fun to build. Simple leave it snapped to the window. You don't have to turn it off or worry about replacing batteries. Just sit back and enjoy free radio reception. In fact, you can even hear those locals on a cloudy day.

Although a regular LED may be used in this transistor tester, a flashing LED was used instead. The flashing LED operates from 3 to 5 volts with 20 mA of current. Not only is the flashing light readily seen, but it gives a little class to the project. Only 12 solar cells and a flashing LED is used in this project.

This little transistor tester will indicate if a transistor is open, leaky, or normal. Simply connect the test lead clips to the base and corresponding terminal of the transistor for the various tests. This tester checks the transistor for quality and indicates the correct npn or pnp type. You may quickly run through several surplus transistors and cull out those defective ones (Fig. 12-1).

Choosing and Mounting the Cells

The small project is mounted inside a 4¼ × 2½ -mini-box. A plastic mini-box was chosen since this type of tester may be knocked around a little on the work bench (Fig. 12-2). Choose a piece of foam to fit snugly inside the plastic box to mount the solar cells on. Cut the foam so when the plastic cover fits over the solar cells, they lay snugly against it. See Table 12-1.

After cutting the foam, drop it down inside to see how it fits. Drill a ¼ inch hole through the middle and side of the plastic box to

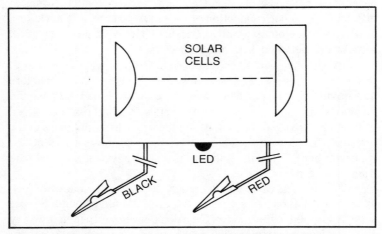

Fig. 12-1. This little flashing transistor tester will indicate if the transistor is open, leaky, or normal.

Table 12-1. Parts List for Project 12.

12 — pieces or crescent solar cells
1 — Flashing LED — Radio Shack 276-036
1 — 4¼ × 2½ × 1½ mini-box.
Misc. — Solder, red and black test leads, two small metal clips, foam, etc.

Construction time — 6 hrs.
Cost of project — $12 to $25, depending if a mini-box or plastic box is in the junk box.

mount the LED. Let the drill go into the foam area for mounting. Now, drill two ⅛ inch holes on each side of the LED for the test leads (Fig. 12-3).

Start laying out the solar cells on the foam area. Choose either pieces of cells or crescent type cells. Any cell with 20 mA or higher rating will do. You may use 12 broken pieces of cells for this project. If broken cells are used, select ones with a heavy bar across the surface area for easy soldering.

Now, lay the pieces of cells in one or two rows as needed. If crescent cells are used, line them up 1/16 inch away from each other. All 12 cells will easily fit on the cut foam area. After laying out the cells, they are ready to be connected.

Wiring the Solar Cells

Before the cells can be mounted, a 1 inch connecting wire should be soldered to the bottom of each cell. Use the soldering iron

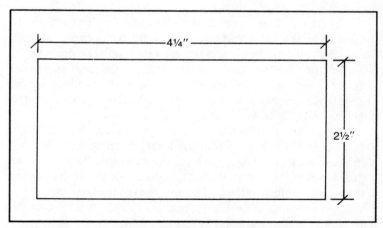

Fig. 12-2. A 4¼ × 2½ × 1½ plastic mini-box was chosen to house the small tester since this type of a tester may get bounced around and receive a few hard knocks.

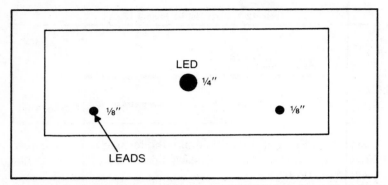

Fig. 12-3. Only three holes are drilled in the side of the plastic mini-box. Use either a drill or the sharp end of a small soldering iron.

lightly on the cell, but make a smooth soldering connection. Very fine wire may be used when connecting the solar cells. Do one cell at a time. Then lay it back in its original position after connecting the small lead.

When crescent cells are used in a straight line, only a small bead of clear silicone rubber cement is needed to hold them in place. If broken pieces of cells are used, place a dab of silicone rubber cement on the bottom of each cell. Try to keep the cells in line as they are placed into position. Go over each cell and level them up. Keep the small connecting wire straight up, out of the silicone cement. Let the cement set up over night or at least for several hours.

After the cells are cemented into position, connect each wire to the next cell. The small lead will go to the top bar of each solar cell. Scrape the cell and place a solder bead with a small soldering iron. Now, pull the wire over to the soldered area. Solder the wire to the next cell. Hold the wire down with a pocketknife or a lead pencil. Connect each cell in the very same manner. All cells are connected in series (Fig. 12-4).

The beginning and end wire connections should be long enough to go to the LED and test lead. Push a hole down through the foam with a small nail or ice pick and bring the two connecting wires out of the bottom of the foam. Connect the positive lead (bottom of cell) to the anode terminal of the LED. Now, the red lead will connect to the cathode terminal of the LED. The black test lead connects to the topside wire of the last solar cell (Fig. 12-5).

Push the test leads through the plastic ⅛ inch holes and tie a knot so they will not pull out. Also, apply silicone rubber cement

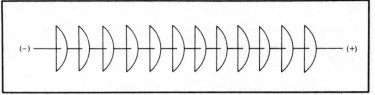

Fig. 12-4. Lay the crescent cells out in one straight line. Connect all cells in series to get the required 3 to 5 volts.

around the knot and the plastic box. This will help to hold the test leads from twisting and breaking off wires to the cells and LED. The base soldered connections do not have to be open since they are down in the foam area.

Testing

Place the transistor tester under a reading light or out in the sunlight. Keep the LED out of the light and towards you, so you can readily see the LED flash on and off. Now, touch the two clip leads together. You should see the red LED flash on and off. If not, double check all connections.

First, measure the voltage at the black lead and at the anode connection of the LED (Fig. 12-6). The voltage should be from 3 to 5

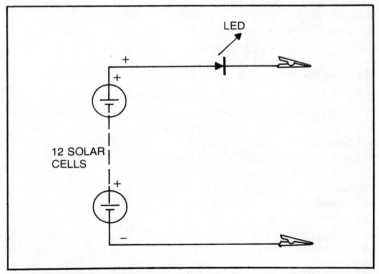

Fig. 12-5. Connect the positive lead (+) to the anode terminal of the LED. The black lead connects to the topside of the last solar cell.

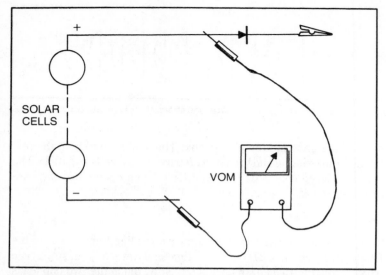

Fig. 12-6. To test the solar cells, measure the voltage at the black lead and anode connection of the LED for 3 to 5 volts. Check the polarity at this point.

volts, if the solar cells are wired up correctly. Now, check the LED connections. Reverse the LED connection, you may have it wired up backwards. Recheck by shorting the two test leads. Place a thin clear plastic lid over the solar cells and secure with four screws which held the metal plate of the mini-box. Okay, now the LED is flashing and it's ready to check out those transistors!

Checking the Transistors

A normal or good transistor will show a flashing light between the base to collector and base to emitter in only one direction. In other words, place the red clip (+) to the base terminal of transistor and the black lead (−) to the collector. If the LED flashes on and off, you may assume the transistor is a npn type and normal (Fig. 12-7). Now, remove the black lead (−) from the collector terminal and clip to the emitter. Flashing LED indicates the transistor is normal. A normal npn transistor will flash the LED with base to collector and base to emitter terminals.

When the red clip (+) is connected to the base and the black lead (−) to the collector and the LED does not flash, the transistor may be open or a pnp type. To determine if the transistor is a pnp type, you can look it up in a transistor book or reverse the red and black leads. Now, connect the black lead to the base terminal and

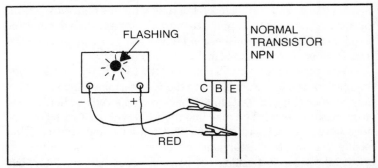

Fig. 12-7. The transistor is normal and a npn type when the LED flashes with the red lead (+) at the base terminal. Use the base terminal for a common junction for all normal tests.

the red lead to the collector. If the LED flashes, you may assume the transistor is a pnp type. Remove the red clip from the collector terminal and transfer it to the emitter terminal. If the LED flashes on and off, the transistor is normal and is a pnp type (Fig. 12-8).

Remember the base terminal is always checked against the collector and emitter terminals for quality tests. Either the black lead (–) or the red lead (+) clips to the base terminal for normal transistor tests. With the transistor out of the circuit, normal, open, and leakage tests can be made.

The Defective Transistor

Since there are more npn type transistors used today in electronic consumer products, always connect the red lead (+) to the

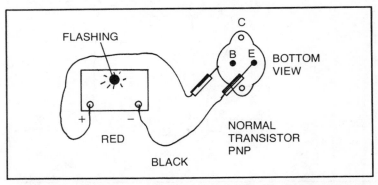

Fig. 12-8. The transistor is normal and a pnp type if the LED flashes with the black lead (–) connected to the base terminal. Remember the base terminal is always checked against the collector and emitter terminals.

base terminal. You may save a lot of time with this test. Now, check for normal npn tests by clipping to the collector and then the emitter terminals. If the LED flashes on both tests, you may assume the transistor is an npn and normal.

When the transistor will not cause the LED to flash with either the black or red lead connected to the base terminal, the transistor may be open (Fig. 12-9). Double check with the red and black leads at the base terminal. You may find the LED will flash with connections between base and collector, but not base and emitter. This means that the transistor junction between base and emitter is open. Likewise, no flashing between the base and collector, but normal flashing between the base and emitter is normal, except the junction between the base and collector is open.

Check the transistor several times between the base and collector and the base and emitter. Reverse the test leads and repeat. If any one test has no indication, the transistor is open between these two elements and should be discarded. But, if the LED flashes with reversed test leads in both tests, between base and another element, suspect a leaky transistor (Fig. 12-10).

You have checked the transistor for open conditions, now let's check it for leaky problems. A transistor is either normal, open, leaky, or intermittent. The leaky or shorted transistor may cause the LED to flash between any two transistor elements.

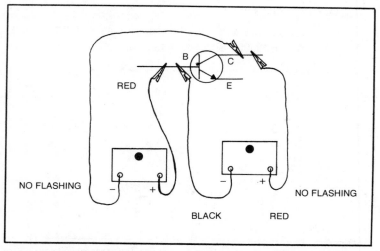

Fig. 12-9. When the transistor will not cause the LED to flash with either the black or red lead connected to the base terminal, the transistor may be open. Double check both leads at the base terminal.

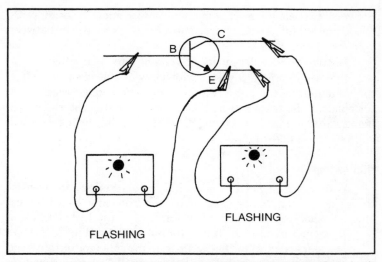

Fig. 12-10. A leaky transistor will make the LED flash with reversed test leads across any two terminals. All transistors should be checked for leaky or shorted conditions.

All transistors should be checked for leaky or shorted conditions. To properly test the transistor for leakage, you will actually make six tests. If the LED flashes between any two terminals with reversed test leads, the transistor is leaky. The transistor may be normal if the LED flashes between base and collector terminals. But, if the LED flashes with reversed test leads on base and collector terminals, the transistor is leaky. Likewise, the transistor may be normal if the LED flashes between base and emitter terminals. Again, if the LED flashes with reverse test leads on base and emitter, the transistor is leaky. The transistor is always leaky when the LED flashes between collector and emitter terminals. You may find the transistor has leakage between two terminals. A shorted transistor may have leakage between all three terminals. A transistor with only a small amount of leakage, may not cause the LED to flash with this type of tester.

Transistor Damage

It's practically impossible to damage a transistor with this type of tester. Although, the supply voltage from the solar cells may be from 3 to 5 volts, the actual voltage across any two elements of a rf transistor is less than .02 V. The voltage between two elements of a low signal of transistor is less than .015 volts. A .04 voltage is found between two elements of a medium power audio transistor. On larger power output transistors, only .05 volts is measured between elements being tested.

Besides the small amount of voltage, the solar cells voltage will go down to practically zero with a low or a dead short. In fact, the voltage from the solar cell may be connected across any two terminals of the transistor and not damage the transistor. Besides being safe while working around the transistor, this little tester will cull out those defective ones.

Npn or Pnp

With this little gadget you can quickly determine if the transistor is a npn or pnp. Simply connect the red lead to the base terminal and the black lead to the collector terminal. If the LED flashes on and off you may assume the transistor is a npn type (Fig. 12-11). Go a step farther - remove the black clip from the collector terminal and clip to the emitter terminal. The LED again will flash off and on if it's a npn type.

In case the LED does not flash off and on when clipped to the emitter terminal, the transistor may be leaky. To prove that the

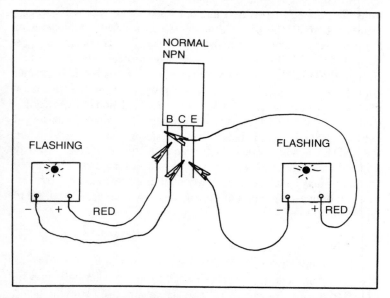

Fig. 12-11. If the LED flashes with the red lead (+) connected to the base terminal and either the emitter or collector terminal, the transistor is a npn type. Reverse the test leads to check for a leaky transistor.

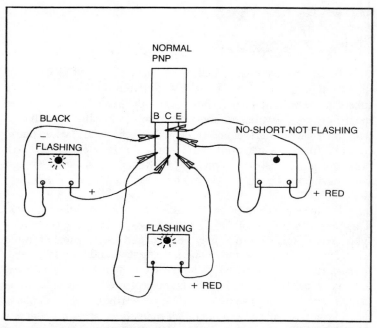

Fig. 12-12. When the LED flashes with the black lead (−) connected to the base terminal and either the emitter or collector terminal, the transistor is a pnp type. You can quickly sort out transistors for npn and pnp types.

transistor is leaky, reverse the red lead and clip it to the collector terminal with the black lead connected to the base terminal. The transistor is leaky if the LED flashes in both directions. Discard the transistor.

To determine if the transistor is a pnp type, connect the black lead to the base terminal. Now, clip the red lead to the collector terminal. If the LED flashes, you may assume the transistor is a pnp type (Fig. 12-12). To make certain, connect the red lead to the emitter terminal and if the transistor is a pnp type with normal operation, the LED will flash on and off.

When the LED flashes with the positive lead (red) connected to the base terminal and the black lead connected to either collector or emitter, the transistor is a npn type. Likewise, when the LED flashes with the negative lead (black) connected to the base terminal and the red lead connected to either the collector or base terminal, the transistor is a pnp type. But, if the LED flashes in both directions with any two terminals, the transistor is leaky between these terminals.

Today, in most homes there is still at least one tube-type radio or TV. In fact, 50 to 60% of the TVs found in the home still have tubes in them. Tubes may be found only in the horizontal and vertical sweep sections of some TV chassis. In other TV chassis you may locate tubes that are wired in a series hookup. When one of the tube filaments goes open, the whole string is out (Fig. 13-1). Now, you must locate the open tube. With this small heater checker, you can check those tubes, very quickly for open heater conditions. See Table 13-1.

To check the condition of the tubes, remove one tube at a time. This prevents the possible mix-up of tubes and putting the wrong tube in a socket. The tube may be damaged when inserted in the wrong socket. Either remove one tube, test it, and then replace it, or mark each tube with masking tape. Be careful in handling tubes as you may break them or rub off the number printed on the side. Don't use a rag to wipe off the tube. You may destroy the tube number.

If you have already removed all tubes for testing without any thought on replacement, check for a tube layout chart on the side of the TV cabinet or chassis. Usually, a white paper tube chart shows where each tube is mounted. In the case of no tube layout chart, check with your local TV technician since he has access to the information about where most tubes go. At least he may look up the make of the set and draw a rough tube layout.

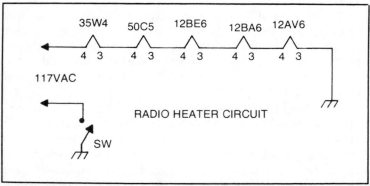

Fig. 13-1. When one tube goes open in a series tube string of a radio or TV chassis, all of the tubes are dark. One of the tube heaters may be open or there is a defective socket or voltage dropping resistor.

Circuit Description

The four most common tube sockets are wired in a parallel series circuit with a power source and LED as indicator (Fig. 13-2). All tube heater terminals are wired in parallel, while the indicating circuit is hooked up in a series-combination circuit with the tube socket. A separate solar voltage source like those found in Projects 1 and 16 is used to light up the LED.

Plug the solar heater checker into the 4.5 or 6 volt socket. Now, remove one tube from the TV chassis and insert it into its appropriate tube socket. If the tube heater is normal or has continuity, the LED will light. This does not mean the tube is good internally, it just means the tube should light up in the TV or radio chassis. No indication may indicate the tube heaters are open. All

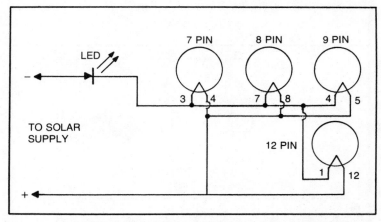

Fig. 13-2. All heaters of the four tube sockets are wired in series with the LED and solar voltage source. When the tube is plugged into the solar heater checker and the LED does not light, suspect a defective or open heater.

tubes are checked in the same manner until the tube with the open heater is found. You may quickly locate the dead tube found in series or parallel circuitry in this manner.

Box Construction

All tube sockets should be mounted on a heavy plastic or metal box. Since the tubes will be plugged in and out, choose a fairly thick cabinet. Here a plastic experimenter's box was chosen to mount the tube socket. Use the bottom side for added strength. Simply lay out the four sockets and mark and cut out holes on the plastic or use a piece of paper taped to the container (Fig. 13-3).

Cut out the holes with a circle cutter, metal steel punch, or soldering iron. A metal circle cutter is ideal to cut out the holes on a metal surface. If a plastic box is used, either the circle cutter or soldering iron may be used to enlarge the socket holes. Always, follow inside the drawn circled areas when using the point of a soldering iron to cut out the plastic.

The plastic holes may be enlarged with the barrel of the soldering iron or pocketknife. Cut or trim off excess plastic on top and underneath each hole. Round out each hole with a rat-tail file. The careful use of a pocketknife may help for the final touch up. Keep trying to fit the socket into the hole for a snug fit.

After each hole has been enlarged, drill the two mounting holes for each socket. Place the correct socket into each hole and bolt it into position with 6/32 bolts and nuts. Mount all tube sockets with the heater terminals at the bottom. Usually, a bolt and nut secure the socket better than metal screws. Metal screws have a tendency

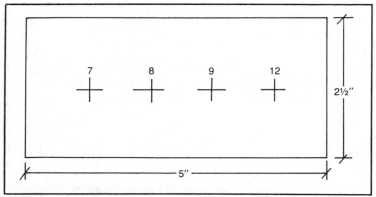

Fig. 13-3. Mark out all four tube socket dimensions on the plastic box. The holes may be cut into the plastic with a metal circle cutter or soldering iron tip.

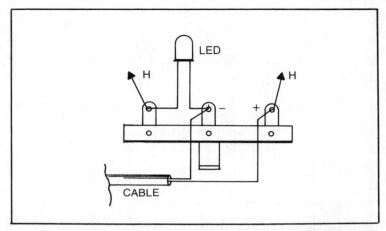

Fig. 13-4. Solder the LED terminal to one outside and the center terminal of the terminal strip. Connect the speaker cable to one outside and the center terminal of the three lug terminal strip.

to loosen, allowing the socket to sag downward. Place a dab of rubber silicone cement over each nut to prevent it from coming loose.

Drill a ¼ inch hole in one end of the plastic case for the speaker or power cable. The point of a soldering iron may be used to make a hole if desired. Be sure to cut off all plastic around the hole. If a metal box is used, place a small rubber grommet in the hole to prevent breakage and shorting out of the power source cable.

Wiring the Unit

Place a three lug terminal strip on one of the tube sockets. Push one end of the speaker cable through the hole and tie a knot in the cable. This will prevent the cable from pulling out and breaking off wires inside the cabinet. Solder the LED terminal to one outside and center common terminal of the terminal strip (Fig. 13-4). Connect one end of the cable to the ground terminal and the positive lead to the open terminal.

Now, each heater socket terminal may be soldered to the two outside terminals of the terminal strip. Mark inside the plastic box where each heater wire is connected. Remember each socket may have a different terminal number. Start at the first left terminal (1) and proceed in a clockwise fashion. Connect wires from each heater pin to the terminal strip. The terminal strip will serve as a tie point for all wired connections.

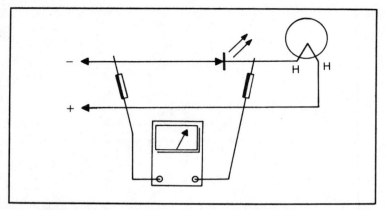

Fig. 13-5. Check for 3 to 6 volts across the LED terminal strip. If the voltage is normal and still the LED does not light, reverse the terminals of the LED.

Testing

When a tube or several tubes do not light up in a radio or TV chassis, suspect a defective tube or component. In radios and TV chassis with a power transformer, suspect a defective tube when only one tube will not light up. If a whole string of tubes do not light up, suspect one tube is open in a series heater string. Check the tube for heater continuity with the solar heater checker.

Plug the solar heater checker into a universal solar power supply. Place the solar cells in the sunlight or under a 100 watt light bulb. Now, insert the suspected tubes. An open tube will not light up the LED. The normal tube will cause the LED to light when the tube is inserted in the correct tube socket.

To check out the solar tube heater checker, insert a normal tube or short across the heater terminals of any tube socket. Simply insert a bare piece of wire or solder down into the heater pins of any socket. Check each tube socket in the very same manner. The LED should light up. If the LED is bright without a tube inserted within the socket, suspect improper wiring of the heater tester.

Double check all wiring connections. The LED should be wired in series with the heater terminals and voltage source. In case the LED does not light with a tube in the socket or with the wire-test, suspect the LED is wired up backwards. Simply reverse the two leads of the LED. Now short out the two heater terminals. The LED should light with voltage applied to the terminals. Check for 3 to 6 volts across the LED terminals (Fig. 13-5).

Chapter 3

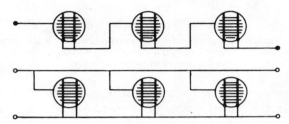

Twelve Solar Projects for Under $35

How would you like to awake each morning when the sun comes up? Well, you may, by just building this simple Solar Sun-Up Alarm. There are no switches or stems to twist, just turn the black box over or point it to the light and it sounds off. To turn it off, simply lay it on the face side (Fig. 14-1).

This solar alarm may be used for a wake-up alarm, or use the alarm to move house plants when the sun strikes a certain area. The alarm may be triggered by the sun hitting a mirror, and reflecting on the solar cells. There are dozens of uses for the solar alarm and it may solve one of your alarm problems.

The solar alarm begins to function when light strikes the solar cells on the front plastic surface. Greater sound is heard as more light appears on the solar cells. The small solar alarm can be used under a regular reading lamp. You will find only a few solar cells and a sounding device component within the black box. See Table 14-1.

A Sonalert® (manufactured by Mallory Corp.) sounds off when a voltage is applied to the terminals. The alarm is a compact solid-state circuit which produces a penetrating sound by purely electronic means with no mechanical contacts. This audible signal

Fig. 14-1. This sun-up alarm consists of only a couple of separate components, solar cells, and a Sonalert® signaling device. The sun-up alarm may be built inside a black plastic box or any other type of container.

device operates on from 1 to 5 volts, using only a few mA of current. Of course, this makes it ideal to be operated from solar cells.

There are four 1 volt solar cells used in this project. If you desire and have them on hand, any solar crescent chips or broken cell pieces may be used to power the sun-up alarm. Simply wire eleven solar cells in series to receive the required 4 volts. This signal device has a 60 dB penetrating noise at 5 Vdc.

Besides the signal device and solar cells, a high impact styrene chassis box was used to house the sun alarm. Actually, only three separate components are found in the solar alarm project. You may mount the solar alarm in any type of cabinet, preferably something that when knocked off a window ledge or desk, will not break in several pieces.

Cabinet Construction

Prepare the high impact styrene box and front panel before any

Table 14-1. Parts List for Project 14.

```
1   — Sonalert (Mallory) SC1.5 — 1TO5Vdc
        #12A2841-1 Gravois Merchandisers, Inc. or at many
        electronic parts stores.
11  — Solar cells — any type crescent, half round, quarter or pieces of solar cells. Here are a few listed below:
        #TM20K187 — H & R Inc. .75 mA
        #S122 — Solar Amp 125 mA
        #P-42,749 — Edmund Scientific Co. (Crescent Cell Mixture).
        #H-12 — Meshna, Inc. 65 mA
        #4924 — Johnson-Smith Co. or use 4 — 1 volt cells
4   — #J801 GE Electronics 1 volt — 50-60mA
                or use,—
2   — #42,710 Edmund Scientific Co. — 3V —3mA
1   — Plastic Chassis box
        #H4-722 GE Electronics
        #158VA —ETCO Electronics
        #270-233 Radio Shack
        #SA 3070-5 — Gravois Merchandisers, Inc.
Misc. — Sheet of plastic for cover, black and clear rubber silicone cement, hook-up wire, etc.
Cost of project — $20 to $30.
```

components are mounted. Remove the four screws and lay the aluminum lid to one side. This lid will be used later to mark and cut out the plastic front panel. Lay the Sonalert® on the center area of the narrow side of the plastic box. Center the hole to be cut with a ruler (Fig. 14-2). Now, lay the Sonalert component on the center area. Draw a circle around it with a pencil.

The 1⅜ inch hole may be cut out with a circle cutter or the sharp point of the soldering iron. It only takes just a few minutes to cut a hole in the styrene case with the soldering iron. First, draw a fine line with the iron on the circle. Each time you go around, push down a little harder. Before long, the plastic circle is ready to drop. Try not to make the hole too much bigger than the pencil line.

Don't worry about the excess plastic building up on top of the case. After the plastic circle drops out, cut off the excess plastic with a pocketknife or file. Check to see if the Sonalert® will fit. If not, use a rat-tail file and go around the circled area. The top part of the Sonalert® component has a lip on it so the whole unit will not come through the case. Clean off the excess plastic inside the plastic box with a pocketknife.

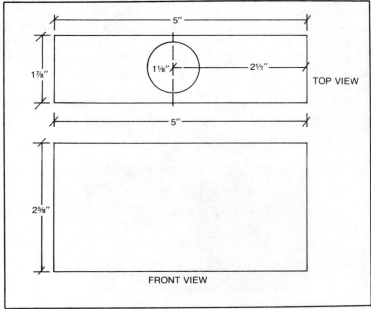

Fig. 14-2. Center the signal device on the top edge and draw a circle around it with a pencil. The 1⅜ inch hole may be cut out with a circle saw or the sharp point of a soldering iron.

After the hole is just right, cement the component in place. Apply a light coat of black rubber cement to the bottom area (Fig. 14-3). Bevel the cement around the signal device. Now, place a layer around the top of the signal device. Bevel the edges with the end of your finger. Wipe up all the excess cement. Not only does the rubber silicone cement cover up any hole defects, but a neat appearance results with a beveled top edge.

Cut the front plastic panel while the black cement is hardening. Select a piece of lightweight clear plastic for the front panel. The thicker the plastic, the less chance of cell breakage when unit is dropped. In fact, the piece of plastic I used was taken from a pre-cut plastic panel used to cover artwork, prints, and photographs. These sheets come in various thickness and sizes. The solar cells will be cemented directly to the panel. Take the metal front cover and lay it on the plastic sheet and mark the outline with a pencil (Fig. 14-4).

Do not remove the protective cellophane covering. This will protect the plastic. Lay the metal box cover on the plastic sheet. Use two straight corner edges of the plastic to start with (which means, you only have to cut two sides). You can cut the plastic with a fine toothed saber saw or with a soldering iron. Cutting with the soldering iron takes only a few minutes and you don't have to worry about cracked edges. Keep a piece of cardboard or wood underneath the piece of plastic so you will not burn the piece of furniture you are working on. Go around the metal piece of the two cutting sides. Apply more pressure each time you go around the metal edges. Hold the metal plate down tight over the piece of plastic.

The four screw mounting holes may be drilled or poked through with the soldering iron. Clean off all rough plastic edges

Fig. 14-3. Apply black silicone rubber cement around the bottom and top area of the sounding device. Bevel the top edge and wipe off the excess cement to make a nice finish.

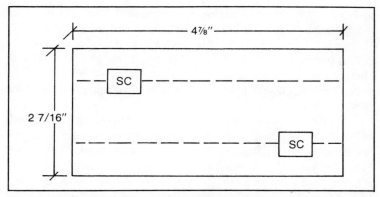

Fig. 14-4. The solar cells will mount on front plastic panel. Cut the panel the same size as the metal front cover.

with a pocketknife or file. If the cutting edges are a little rough run each edge over a flat piece of sandpaper. Be careful, make several passes over the sandpaper to square up the plastic edges. Check to see if the front plastic piece and holes line up properly. The solar cells are mounted on this piece of plastic.

Mounting the Solar Cells

After choosing the desired solar cells, line them up on the clear sheet of plastic. Here four 1 volt solar cells were used and are very easily mounted. If eleven different solar cells are chosen, solder a 1½ inch connecting wire to the front side (– bar) of the solar cell. Since the cells are going to be mounted face down on the plastic, all cells must have a connecting wire soldered to them (Fig. 14-5). The

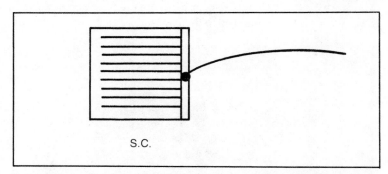

S.C.

Fig. 14-5. When eleven solar cells are used, solder a 1½ inch connecting wire to the face or bar (–) of the cell before the cell is cemented into position. Connect the cells in series.

back side (+) lead can be soldered later or if in line, simply solder the negative lead directly to the back soldered area of the cell.

Apply a drop of clear rubber cement to the front edge of the cell. If longer solar cells are used, place a drop of cement at both ends of the cell. On smaller solar chips, only a drop of cement is needed to hold the cell in place. Do not use black or any other color cement.

Now, lay each cell in its place or in a neat row. Try to keep each cell the same distance apart. You may have to turn a cell or two to one side to get them all in, if larger cells are used. Press down lightly on the cell back area so the cells will stick to the front plastic piece. Set the solar panel to one side and let it setup overnight (Fig. 14-6).

Connecting the Cells

Connect all solar cells in series to get the correct voltage. Take the connecting wire from the front side of one cell and connect to the back side (+) of the next cell. Start at one end and go to the next cell. The beginning and last cell should have a long 6 inch wire connected to the cell area so they can be wired to the signal device.

The black and red wires should be connected together on solar cells with attached leads. Usually, the wires are long enough on the first and last cell to connect to the Sonalert®. All cells with attached leads should be arranged so that the wires will connect to the next cell. In other words, the black wire should connect to the red wire at the top of the next cell. Likewise, the red wire should connect to the black wire at the bottom of the solar cell.

When soldering around the plastic front piece, be careful not to drop any excess solder on the plastic to ruin the front appearance. If the cell wire is to be soldered to the backside of each cell, be careful

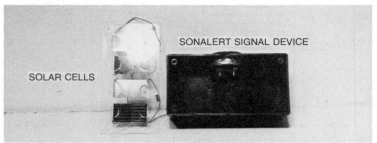

Fig. 14-6. After the solar cells and signal device are mounted let them setup over night.

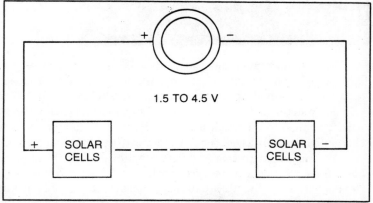

Fig. 14-7. The sun-up alarm is very simple to hookup. Connect all cells in series with the sounding device.

not to apply too much heat or the cement will come loose. To prevent this, solder a lead to each side of the cell before mounting. Cut all connecting leads as short as possible.

Before mounting and wiring the solar panel to the signal device, check for correct voltage with the vom. You should measure between 4 and 5 volts under a strong reading lamp or sunlight. If the voltage is less, check all soldered connections. Now connect the solar panel to the Sonalert® device (Fig. 14-7). Solder the red lead (+) to the positive lead of the signal device. Both positive (+) and negative (−) terminals are marked on the Sonalert®.

Testing the Sun-Up Alarm

To check out the sun-up alarm, place it in the sun or under a reading lamp. The brighter the light, the greater the sound coming from the alarm. The alarm may be operated with the solar cells straight up or with the alarm lying on its side. To shut the alarm off, just turn it over and place the cell side face down.

If the alarm does not sound, suspect a broken cell or connecting wire. Remove the four screws and measure the voltage at the signal device terminals. Place a lamp close by or place the unit under the sun. Check the cells for possible breakage. Usually, the cells remain intact unless rough handling or connecting wires are pulled off the cell surface. When voltage is applied to the Sonalert® and no sound, suspect a defective signaling device. Keep the alarm turned face down when not in use. One thing for certain, this alarm will operate for years without any mechanical problems.

These solar cells and panels may be used in a variety of applications from charging small batteries in radios, calculators, and flashlights to providing power to unmanned radio transmitters and telemetry stations. As long as the sun shines, a small trickle charge will be applied to the small batteries. Here is a duo-solar battery charger which will charge one AA and C battery at the same time.

You may want to charge different batteries than those found here. If so, just add a correct battery holder for any 1.5 V battery. Remember, these circuits are to charge small batteries only. In this project we find two small solar panels (one commercial and the other homemade) for charging each different battery (Fig. 15-1.) Built-in diodes prevent batteries from discharging at night through the solar cells. You may want to construct your own solar voltage panels out of separate cells. The charging voltage must be greater than the voltage of the battery.

Here, we find a solar cell microgenerator panel mounted to the right (Fig. 15-1). The voltage output under normal sunlight is 2.25 volts at 40 mA. Over 2 volts is applied to the charging battery through a series diode. Each battery must have a separate solar panel and diode for correct trickle charging. See Table 15-1.

Making Your Own Solar Panel

Select solar cells with over 36 mA of current capability when building your own solar panel for small battery charging. You will

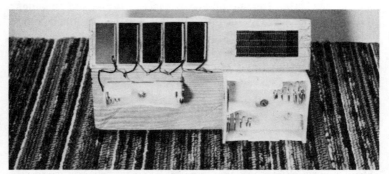

Fig. 15-1. In this photo we see two different solar cell charging circuits. The one to the left is a homemade solar cell unit and the right one is a commercial solar microgenerator.

Table 15-1. Parts List for Project 15.

1 — Solar Cell Microgenerator Kit #42,466 — Edmund Scientific (includes diodes)
1 — 1 or 2.5 amp silicon diode (found at most electronic stores)
5 — Solar Cells — 36 mA or higher #30,735 — Edmund Scientific
     #S122 Solar Amp
     #22A21500-8 — Barsten Applebee
     #SS-10-6LC — Silicon Sensors
     #24-800 GC Electronics
1 — "AA" penlight battery holder — Radio Shack 270-382 or equivalent
1 — "C" cell battery holder, Radio Shack 270-385 or equivalent
Misc. — Mounting wood screws, hookup wire, solder, etc.

need five separate solar cells. After choosing the correct type of cells, place them on a piece of paper and sketch out a plan for mounting. Instead of a 2 × 4 inch board for mounting, you may want to use a 2 × 6 inch or longer piece of wood. Of course, these solar cells may be mounted on plastic or other types of mounting surfaces (Fig. 15-2).

Connect the solar cells in series (Fig. 15-3). Keep the wires short so the cells may be mounted close together. Solder all connections and then place the five cells on the wood surface. Leave the negative (black) and positive (red) leads long so the solar panel can be connected to the diode and solar battery holder. Solar cells may be connected in a series-parallel arrangement to produce the desired power output needed to charge the batteries.

Cement each solar cell to the board with clear silicone cement. Colorless cement does not smear or mark up the mounting compo-

Fig. 15-2. Mount the solar cells directly on the wood or plastic. Choose a piece of flexible plastic or even cardboard and glue it to the wooden mount.

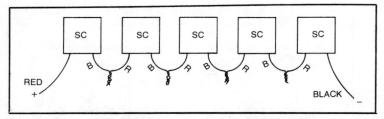

Fig. 15-3. Connect all solar cells in series. Solder the red wire to the black until all cells are connected. If no connecting wires are available, connect the underside wire to the top side wire of the next cells.

nents as easily as black or white. You may want to mount these solar cells on a separate piece of plastic or cardboard and then screw the panel to the wood base. Whatever, weight down the cells so they will stay in place and let them set up overnight.

Now, measure the voltage across the solar panel. Under bright sun or a sunlamp you should have about 2.25 volts. If not, check each connection for a poorly soldered joint. Measure the voltage across two cells at the same time to determine where the poor connection is located (Fig. 15-4).

The Charging Circuit

Almost any silicon diode can be used as a blocking diode. When a diode is placed in series with the charging cell, the battery will not

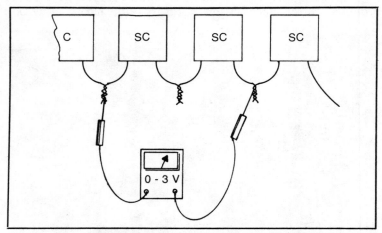

Fig. 15-4. Use a vom and measure the voltage across two different cells to determine if a poor connection exists. You should measure close to 1 volt under sunlight or a strong spotlight.

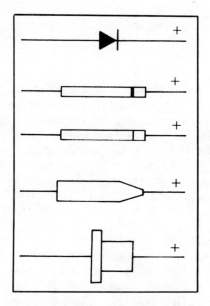

Fig. 15-5. Silicon diodes come in various shapes and sizes. Check the diode for the proper polarity markings.

discharge back through the solar cell panel. Choose a silicon diode rated above 50 mA. Regular 1 or 2.5 amp silicon diodes are ideal and readily available at most radio electronics stores. Remember, the positive lead (+) of the diode must go to the positive connection of the battery holder. The positive terminal of various type diodes are marked as found in Fig. 15-5.

Since two separate batteries are to be charged with separate holders, a different wiring circuit is found with each charging unit. Simply connect each component in series (Fig. 15-6). Each battery holder will be in series with its own circuitry. Here we have two

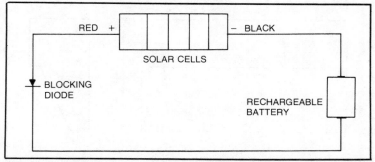

Fig. 15-6. Here is the battery charging circuit. Simply connect each component in series.

125

different charging circuits, one homemade and the other a commercial unit.

Preparing the Base

We choose a 6 ½ inch piece of 2 × 4 to mount all of the components on. This piece of wood was selected since it will fit on most window ledges and it is heavy enough so it will not easily get knocked off. Besides that, it's difficult to break if dislodged from the window ledge. Select a clean piece of 2 × 4 and cut a slot angle at least two inches wide (Fig. 15-7). If a bench saw is not handy, you can mount all components on the whole piece of wood.

Sand down all sides after the piece of wood is cut to length with a cut at the correct angle. Round the corners off if desired. Use medium sandpaper at first and finish up with fine sandpaper for a good finish. Clean off all dirt and smudge marks.

Now, spray the piece of wood with enamel paint or a clear varnish. We choose a clear finish of plastic spray. Of course, clear varnish, Deft, or lacquer spray may be used instead. Usually, clear spray will dry within a half hour. Let it set one hour or so and then go over the entire surface with fine steel wool. Clean off all particles with a cloth. Spray on several coats of clear or satin finish. Let the piece of wood set up overnight before mounting the various components.

Mounting the Parts

After finishing the piece of wood, mount the two battery holders. Only one wood screw is needed to hold each plastic holder in

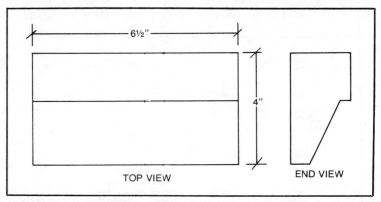

TOP VIEW   END VIEW

Fig. 15-7. Select a piece of 2×4 6½ inches long. If a bench saw is handy, make the angle cut to mount the battery holder.

126

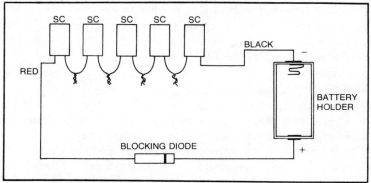

Fig. 15-8. This is a pictorial to help you connect each component.

place. If mounting holes are not already provided, drill or use the small point of the soldering iron to make mounting holes. The size of the battery holders will determine how much mounting space is needed. Place both battery holders on the slanted surface.

Mount the solar cells and blocking diode last. With these two separate solar cell panels, only a drop of silicone cement under each end is needed. If desired, place a piece of plastic over the solar cells for protection. Let the cement set-up before attempting to wire the solar battery chargers.

Connecting the Wires

Wire each solar section as shown in Fig. 15-8. Place the fixed diode between the positive wire and connection of the battery holder. The positive lead of the diode should go to the positive battery terminal. This prevents the battery from discharging at anytime through the solar cell. Remember the red lead of the solar cells goes to the negative terminal of the silicon diode. Actually, the diode is wired in series between the solar cell and battery holder.

Check the battery holder for marked (+ or −) terminals. Look down inside the holder area. If not marked, the battery spring is always negative. This spring fits tightly against the bottom of the "AA" or "C" cells. Some battery terminals may have two wire leads coming from the battery holder. The red wire is always positive and the black or blue is negative. The two separate charging solar cells are wired up to their respective components. Solder one end of the diode to the positive terminal of the battery holder. The other end may be soldered to the long solar wire. Use a piece of spaghetti insulation over the connection or coil the excessive wire to take up the extra slack and make a nice looking appearance. Use regular insulated hookup wire to connect the other components. If the battery holders have self-connected wires, the excess may be used to connect or wire up the other parts.

Checking the Charger

Measure the output voltage across the battery holder terminals, after all wiring is completed (Fig. 15-9). A voltage measurement with the vom will not only show the voltage output but indicate if everything is wired correctly. When the diode is wired backwards, you will not measure any voltage at the battery terminals.

Visually, inspect the diode marked terminals for correct hookup. The white or marked area at one end of the diode is positive and should be connected to the positive terminal of the battery holder. To make sure all soldered connections are proper, measure the voltage across the solar cells. Simply place a reading lamp close to the cells for these tests. Place the red lead of the vom to the terminal from the bottom side of the solar cell, the black lead to the top side of the solar cell.

The voltage should be over 2 volts. Now place the red lead on the positive lead or lead connected to the battery holder. If there is no voltage reading, the diode is wired up backwards. Remove and exchange the ends of the diode. Solder the diode in place and take another voltage reading. Usually, these diodes will not drop the voltage over a ¼ of a volt when wired in series.

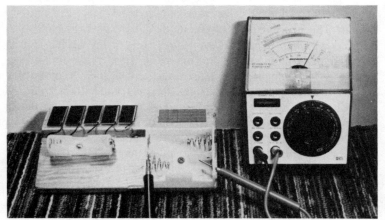

Fig. 15-9. To check out the battery charger, simply measure the voltage across the battery terminals. Under strong sunlight, the voltage should be around 2.25 volts. If it is lower check for poorly soldered connections.

Now, insert the discharged "AA" or "C" cell and set the solar battery charger on the window ledge under the sun. How long does it take to charge up the battery? Of course, this depends upon how low or discharged the cell may be. It's possible with some cells that they will never charge up. But, after a few days, measure the charged cell for correct voltage (1.5 V). Always remove the cell from the charging holder for this purpose.

PROJECT 16
UNIVERSAL SOLAR POWER-PAK

Although, you can now purchase a multi-voltage solar panel for operating transistor radios, toys, motors, and any other small electronic device (with a low current rating) we decided to build one. Most multi-voltage supplies have a 3, 6, and 9 Vdc at 50 mA or less. Here is a solar panel with 3, 4, 5, 5.5, 6, and 9 Vdc with a current capacity of 100 mA (Fig. 16-1). You can construct this solar panel for about $20 or $30. Of course, you may not save a lot of money by building one yourself, but it's lots of fun and you can see firsthand how the solar cells perform. See Table 16-1.

How to Begin

The universal solar-pak was built inside a 5 × 7 plastic picture frame. Originally, the picture frame was designed to hold a 5 × 7 enlargement. If a plastic box is not handy, these frames can be picked up at the photo counter in any retail store. These clear

Fig. 16-1. Here is a photo of the completed universal Solar Power-Pak. You may select any voltage between 3, 4.5, 5.5, 6, and 9 volts.

Table 16-1. Parts List for Project 16.

Solar Cells —
        24 at 100 mA or better
        12 — #5124 — Cut into to make 24 Solar Amp.
        24 #H-11 John Meshna, Inc.
        24 #42,268 Edmund Scientific Co.
        24 #TM 21K666 — H & H Inc.
              or, —
        12 #TM 21K932 — H & H Inc. Cut into to make 24 cells
6 — Banana jacks
        6 — #274-725 Radio Shack
        6 — #33-210 tip jacks (red) GE Electronics
        6 — #12-01049 -2-002-6 tip jacks, Gravois Merchandiser
        6 — #100ZK ETCO Electronics
2 — Banana Plugs
        2 — #274-721 Radio-Shack
        2 — #33-106 red tip plugs, GE Electronics
        2 — #12-01050-2-002-6 tip plug Gravois Merchandiser
Misc.
        1 — 5×7 plastic photo frame, hookup wire, solder and
            clear rubber silicon cement.
Commercial Universal Solar Panels
        #S116 — 3V, 6V and 9 solar amp
        #42,955 — 3,6 and 9V, Edmund Scientific
        #S11-104 — 3,6 and 9V — H & R Inc.
Cost of Project — $20 to $30.
Construction Time — 4 to 6 hours.

plastic frames are very sturdy and will protect the solar cells from possible breakage.

A piece of ⅜ inch foam was used to mount the solar cells (Fig. 16-2). These foam panels can be located in most hobby stores. Simply lay the plastic frame on the piece of foam and cut along the edges with a pocketknife or razor blade. Cut the foam so it will just fit down inside the plastic frame. Make sure the piece of foam fits down inside the plastic frame before mounting the solar cells.

Selecting the Solar Cells

There are many types of solar cells on the market that may be used in the multi-voltage solar panel. If you desire to build only a 50 mA panel, select solar cells with that current capacity. For a 100 mA or larger current supply, we selected twelve .225 amp crescent cells and cut them in half. These cells are ¼ of a 3 inch solar cell. If you desire a solar panel with a greater current rating, select solar cells of 150 or 250 mA.

The twelve .225 amp cells were cut in half making a total of 24. Actually, 24 solar cells will give a total of over 12 volts without a load, which will provide a good working voltage of about 9 volts. Either select 24 solar cells of the desired current rating or cut the 12 cells in half.

Although it may not be wise to try to cut or break solar cells, you can see from Fig. 16-1 that these twenty-four cells turned out quite well. We must remember these solar cells were crescent or ¼ inch sections of original round cells. Do not try to cut or break whole or half round cells. It's too easy to break and ruin expensive solar cells.

Cutting the Cells

Take one cell at a time and place it on a solid surface such as a board, desk, or table for cutting. Place the cell face down so the solid solder or silver side is up. Now, draw a pencil line down through the middle of each cell. Center the line on each cell. Take a ruler and lay it upon the line across the cell. Use a razor blade or a sharp blade of a pocketknife (Fig. 16-3) to cut the cell.

Rake the blade across the cell. Do this several times to start a thin groove so the cell will break at this cut. Now, place more pressure on the knife until the cell breaks. Don't look for a clean cut edge. Don't be in a hurry. If you are careful, you can easily have 24 working solar cells.

If the cells are a little ragged, don't fret. But, if the cells break into several pieces, then you're in trouble. When the cell breaks into little pieces, save all of them for future solar projects that operate on lower current. With patience and care you now have 24 separate solar cells for the price of twelve. You may want to choose some solar cells that do not have to be cut.

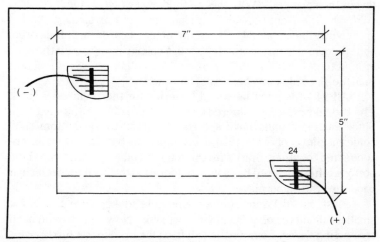

Fig. 16-2. Here is a solar cell layout. Place the cells in the 5×7 inch space.

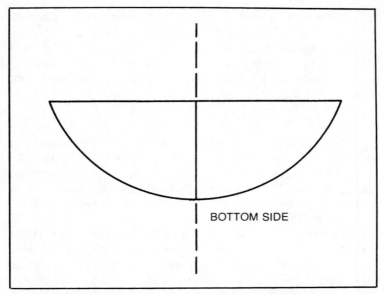

BOTTOM SIDE

Fig. 16-3. The larger crescent or ¼ inch cells may be cut. Lay a ruler on the middle section of the solar cell. Draw a cutting line alongside it with a razor blade or pocketknife.

Preparing the Cells

Before the cells can be wired in series, place them on the cut piece of foam. Better yet, cut a piece of cardboard the same size to layout the solar cells. With this method, you won't accidentally drop solder on the foam (it will melt right through) or damage the top area. Now, line up all 24 cells on the piece of cardboard. If desired, shift them into a line or other pattern. You may have to shift the cells around to get them all in this space. The cells may be reversed so that the flat sides are next to each other.

All 24 cells must be wired in series for the required 9 volts. The topside must be soldered to the bottom side of the next cell. Since each cell must have a wire connected to the bottom side, solder a piece of wire to the bottom side before any cells are connected together. Start with the top left cell and work toward the bottom. After each cell has a wire soldered to the bottom, set it right back into the same place.

First, cut 23 1½ inch pieces of flexible hookup wire. Select the smallest diameter of hookup wire available. Now, tin one end of all connecting wires. Remove the cell from the line-up and turn it over. Solder the small piece of wire to the bottom cell area (Fig. 16-4).

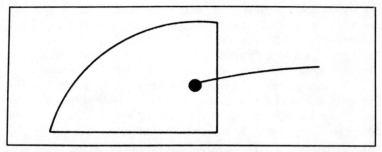

Fig. 16-4. Solder a ½-inch connecting wire to the bottom of each cell. After mounting, the cells can be wired in place.

Try to solder the wire at the end near the next cell. Do not leave the iron on too long, but let the solder flow. Keep all high points of solder at a minimum. Bend the wire upward and place it back into the line-up. Solder a piece of connecting wire to each cell before mounting.

The solar cells can now be mounted on the foam piece. Lift each cell (one at a time) off the cardboard and transfer to the foam panel. Place a large drop of clear silicone rubber cement to the center bottom area of each cell. Mount the cells in line or the same layout as on the piece of cardboard. Press the cells down into the foam panel. Don't push too hard or you might damage the cells. Let the cement set up overnight.

Connecting the Cells

After the cells are cemented in position, they can be connected together. Make sure the bottom (+) lead goes to the top (−) connection of the next cell. Connect the lead to the center bar (−) of each cell area (Fig. 16-5). Cut the flexible connecting wire just long

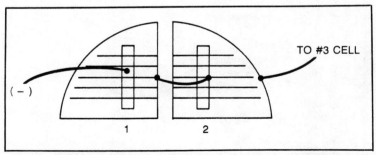

Fig. 16-5. Connect all cells in series. Cut the connecting wire just long enough to solder to the middle bar (−) of the next cell.

enough to solder to the center bar of the next cell with a pair of side cutters. Tin the end of the connecting wire and solder to the bar area. With some types of flexible hookup wire, the insulation may come off or shrink back without having to scrape off the insulation. Try this method to prevent pulling a cell from the foam panel.

Be very careful not to apply too much heat from the iron. It's possible to unseat the cement on the bottom side of the cell. Too much heat may lift the solder from the cell's top surface. Keep all sharp soldering points at a minimum. Remember this topside fits snugly against the plastic frame. Do a very neat job since this side can be seen from the top.

When soldering, apply solder to the wire lead and bar area. Hold the wire down with a pencil or a pocketknife blade. Use only a small amount of solder. Too much solder is difficult to smooth out and makes a large connection. Start at the top left hand corner and go to the next cell until all 24 cells are connected.

If one of the cells comes loose from the foam panel in the soldering process, turn the cell over. Apply a dab of silicone cement and start again. Hold the cell down with the point of a lead pencil. You may sneak some cement underneath the cell with one edge raised up. Only use clear rubber silicone cement. This clear cement will not show or smear like black or grey cement against the white foam panel (Fig. 16-6).

The last bottom cell should have a six inch connecting wire soldered to it. Likewise, solder a six inch piece of flexible wire to the top or beginning cell (−). The negative or topside cell wire will be soldered to a silicon blocking diode. Take a voltage check after all cells are wired together.

Checking The Voltage

Place a 100 watt lamp nearby if the sun rays are not available.

Fig. 16-6. Here is the finished solar panel constructed on a ⅜ inch piece of foam. Check at each tap connection for the required voltage.

134

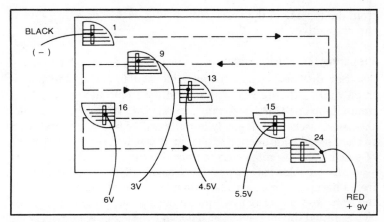

Fig. 16-7. This is a schematic showing how the cells are connected with color coded wires. Notice the various voltage taps.

Clip the black lead (−) of the vom to the top cell connecting wire. Connect the red vom to the remaining solar wire (+). The vom should measure at least 10 volts dc. If not, recheck each cell connection. Check for a poor or broken soldered connection. You should have no problem if all of the connecting wires were soldered carefully at the beginning.

You may want to start at the top left cell (−) and measure the voltage across the two cells. Keep the vom set at the lowest dc range. Go from cell to cell until you have located a poor connection or broken solar cell. Under direct sunlight you should have a voltage over 12 Vdc. If a broken cell is found, remove and replace it.

After the correct total voltage is measured, you can start tapping off the various voltages (3, 4.5, 5, 5.5, 6, etc.). Count off 8 cells from the negative end (1) and poke a hole through the foam panel with the ice pick or similar instrument (Fig. 16-7). Be careful not to break the foam in half. The 3 volt wire will be connected to the top side (−) of the 9th solar cell. Select a flexible piece of hookup wire and poke it through the small hole. Solder the wire in place and pull the excess wire down through the foam. Likewise, make a top or voltage connection at the 12th, 14th, and 16th solar cell. Later these connecting tap wires will be connected to their respective voltage jacks on the plastic frame. You may want to use a different colored wire for each voltage tap connection. For instance, the negative (−) lead would have a black wire and the positive (+) or 9 lead would be a red colored wire. The color code may be followed as shown in Fig. 16-7. Tin the ends of each wire and poke it through from the topside of the foam. Now solder each connecting wire to the respective solar cell.

Preparing the Plastic Frame

Set the solar panel to one side and finish up the plastic framework. Drill six ¼ inch holes on the front side of the plastic frame. Line up the holes so they are balanced and even. Make sure the banana jacks will fit easily into the drilled holes. Be careful not to marr the topside section of plastic. It may be best to tape a piece of paper over the remaining top panel to prevent scratch marks.

The six holes may be pushed through with the tip of a soldering iron. Line up the holes and mark where each jack goes with the tip of the iron. Now push the iron tip into the plastic, enlarging the hole. You will find the hole must be a lot bigger since the plastic, while hot, will shrink over the hole. Cut off the excess plastic with the pocketknife. If the holes are not big enough to place the jacks through, simply insert the iron tip and enlarge each hole. Keep the excess plastic cut off as the jacks will not mount flush with plastic around the hole area. Do not mount the jacks until the front solar panel is placed into position (Fig. 16-8).

The Final Touches

Now, place the solar foam panel inside the plastic frame. Insert all jacks and tighten in place. Solder all cell wires to their respective jacks. Connect the silicon diode between the negative lead and jack (Fig. 16-9). Check each jack for the correct voltage. You should have

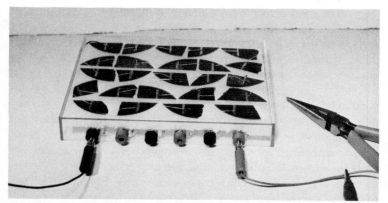

Fig. 16-8. Mount the banana jacks in the front of the Solar-Pak after the front solar panel is placed inside the framework. Connect the required voltage taps to each jack.

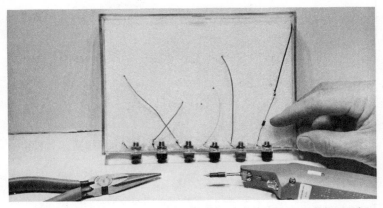

Fig. 16-9. Here is how the bottom view of the solar pak looks after complete wiring. Notice the silicon diode is inserted in series with the black lead (−).

from 1 to 2 volts more than needed at each jack. But, when loaded down, the voltage taps should be about right.

Cut a piece of ⅜ inch foam and make a tight fit over the wires and jack. Before placing the bottom piece into position, apply clear rubber silicone cement down around the jack areas. Fill this area full so a nice finish will show through the plastic. Now cement the bottom foam section. Seal with silicone cement around all edges to keep dust and dirt out of the cell area. Label each jack with the correct voltage.

Remove the batteries from your favorite portable radio. Clip the leads from the universal power-pak to the radio battery terminals. Check the number of batteries to determine the correct voltage. If 4 "C" cells are used, the correct voltage is 6 volts. When one single 9 volt battery is used, plug in the jacks to the two outside terminals. It doesn't matter if you clip the wrong leads on to the various battery terminals. You can't hurt anything with a protection diode in the circuit. Simply reverse the solar-pak leads and sit back to happy music. In fact, you never have to worry about buying batteries for the radio again!

PROJECT 17
ADD SOLAR POWER TO YOUR PORTABLE RADIO

How about solarizing that small portable radio? You can purchase solar panels that can be glued to a portable radio. Usually, there is the only room available to mount the solar cell, since the

Fig. 17-1. This is a commercial solar panel connected to a portable Sanyo radio. These panels come in 6 and 9 volt types and range in price from $12 to about $20.

topside of the radio may be taken up with the carrying strap or antenna (Fig. 17-1). You may want the flat solar panel to lie alongside, but connected to the radio. See Table 17-1.

Of course, every small portable radio can not be solarized unless the solar unit is constructed in a separate power panel. Most small radios pull less than 50 mA of current. There are 30 mA, 6 and 9 volt solar panels on the market for portable radios. They actually replace the self contained 6 or 9 volt battery. These small solar panels will not operate radios requiring larger amounts of current.

Mounting the Solar Panel

Select a flat place on the back of the portable radio to mount the solar panel. Keep the panel away from earphone jacks and AM-FM switches (Fig. 17-2). If there is not enough room to mount the solar panel, simply connect it and let it lie beside the radio.

Before fastening the solar panel to the radio case, make certain that the plug and cable are long enough to go inside the battery compartment. If not, simply cut the cable in two, insert a couple of

Table 17-1. Parts List for Project 17.

Commercial Solar Panels $12.00 to $19.00
S100 — 9Vdc 30mA Solar panel — Solar Amp, Inc. $12.00
S116 — Universal, 3, 6, 9Vdc Multivoltage panel Solar Amp, Inc. $18.95
42,955 — Universal 3, 6, 9Vdc — Multi-volt panel — Edmund Scientific Co. and other sources $19.95
Time to mount — 2 hrs.
Cost of constructing your own — $18.00 to $25.00
#30,735 — 36mA $3.95 each — Edmund Scientific Co.
S122 — .125 amp — 4 for $6.00. Solar Amp Inc.
Construction time 4 to 6 hrs.

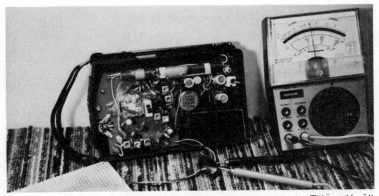

Fig. 17-2. The solar panel clips directly to the battery terminals. Either place it alongside the radio or mount it on the backside.

lengths of wire and solder up. When you finish, make sure the red wire is connected to the same red wire as in the beginning and the black to the black wire. Solder and tape up the two leads.

Either drill a new hole or cut a slot along the edge of the back panel. Use a small tip of the soldering iron to drill a hole in the plastic cover or take a pocketknife and slot out the plastic edge. If you must drill a small hole in the plastic case, the wires must be cut in two so the cable or two connecting wires can be soldered to the battery terminals. Always, solder and tape up those bare connections.

After the cable is in place, select the spot where the solar panel will mount. Apply a drop of rubber silicone cement on each corner of the backside of the solar panel. Now press the solar panel in place. Use either black or clear silicone cement for this purpose, since either color will match the solar panel or radio. Place a book or heavy object on the panel to hold it in place. Allow it to set up overnight.

With the panel securely glued to the back panel, connect or plug in the battery connections. Now the radio is ready to operate with the solar panel. In some portable radios, the solar panel may operate under the reading light. But, actually solar cells were designed to operate in the sun. Just place the backside of the portable radio into the sun and it will operate for years (Fig. 17-3).

How to Roll Your Own

You can design and construct your own solar panel out of separate solar cells. First, determine how much current the small

Fig. 17-3. Here the small solar panel is mounted to the rear cover. Select a spot where switches and earphone jacks are out of the way. Place a drop of silicone cement in each corner of the solar panel.

portable radio pulls and what is the operating voltage. Sometimes, this data is included in the service manual contained with the portable radio. If not handy, figure it out yourself.

The operating voltage is easy to handle. For instance, if four penlight cells are used to power the small radio, the operating voltage is 4 × 1.5 volts or a total of 6 volts. Each penlight battery is rated at 1.5 volts. If only one battery is found, look on the side of the battery for a 9 volt listing.

Now to check the current needed to operate at the correct operating voltage. Take a small vom and turn the range to 300 mA. In radios with a battery plug, just plug in to the side of the battery and let the other terminal angle out over to one side (Fig. 17-4). Place the vom current meter leads in series with the battery and plug. It doesn't make any difference which set of plugs you use.

Turn the radio switch on. If the meter hand goes backward (goes to the left without a reading) reverse the two meter leads. Now, read the actual current the radio is pulling. You may want to turn the current range down to a smaller current reading. Check the current with the radio off-station. Now, turn the radio on to a local station. You may find the current will almost double when the volume is turned up. Take the highest current reading.

With this Sanyo AM-FM radio, the current and voltage needed to operate with a solar panel was 6 volts at 30 mA. With the volume down, the current was 20 mA and when with regular listening volume, the current was around 30 mA. Now, we need to design a 6 volt solar panel at 30 mA of current.

To construct a solar panel of 9 volts at 30 mA of current, connect 21 cells in series. Select cells with a crescent or ¼ round shape. Usually, they are portions of a 2¼ or 3 inch solar cell. Be sure and check the current rating of the cells. They should be 30 mA or higher.

Selecting the Solar Cells

It's always best to select solar cells with a current carrying capacity a little greater than the current needed. For instance, a 50 mA 6 volt solar panel will operate the portable radio without pulling down the voltage of the 30 mA types. So select solar cells with 30 mA of current or larger. Always remember the larger the current the greater the size of the solar cells. The solar cells must mount in the required space for the needed current and voltage rating of the solar panel.

If you want to design a 6 volt 30 mA solar panel to operate this particular radio, choose available solar cells of 36 mA or higher in current rating. Practically all solar cells have an operating voltage of .45 volts. Now, to construct the solar panel, obtain 14 solar cells rated at 36 mA or higher.

Connecting the Cells

Place the fourteen cells on a piece of paper and arrange the cells so they can be connected in series and will occupy the smallest space (Fig. 17-5). These small cells may be rectangular, round,

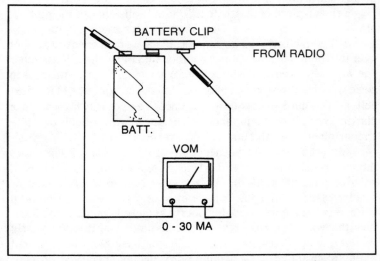

Fig. 17-4. How to connect the milliampere range of the vom to the self-contained battery and radio. Take the highest current reading with the radio operating.

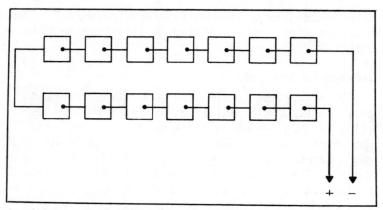

Fig. 17-5. Lay the 14 solar cells out on a flat piece of paper or cardboard to determine size of solar panel and proper arrangement of the cells.

quarter, or crescent. Whatever, connect the cells in series to acquire the needed 6 volts. Use flexible wire to connect the various cells.

It's best to solder a wire to the bottom side and a connecting wire to the topside of the cell. If the cells have two large silver lines, solder the wire to both lines for greater current capacity. Do this to all cells before mounting them. The leads should not be over two inches since the cells will be mounted closely together.

Select a piece of plastic to mount the cells on. Cut the piece of plastic to the correct size. Now start at one corner and mount the first cell with clear silicone cement. Place a small speaker magnet over it to hold it in place or use another suitable weight. Remember one wire will be on top of the cell and one under. To place the cells in series, connect the topside of one cell to the bottom side of the next cell. Do this until all of cells are in place. Temporarily lay a piece of plastic over all the cells and weight down with a book or some similar object. Let the unit set up overnight.

After the cells have been mounted, measure the voltage across the two remaining wire ends. Under sunlight you should have 6 volts or more (Fig. 17-6). When checked under a fluorescent or regular reading light, the voltage may be only 3 volts. If the total voltage is very low, check for poorly soldered connections. Measure the voltage across two solar cells until you find the poorly soldered joint. Repair and replace. If you have been very careful when soldering each connection no problems should be encountered.

142

Finishing the Panel

Place a ⅛ inch strip of plastic around the top edge next to the solar cells. This will support the top plastic panel. To prevent damage to the cells, and if used outside, the cells should be protected from various kinds of weather. If it's going to be used only inside and out of the weather, the top plastic piece is not necessary.

Now, use plastic liquid or airplane glue and seal the strips to the solar panel. Select a thin sheet of plastic, or if available, use clear plastic with lens action material. This lens piece is plastic with bubbles or sharp indents to create lens action on the solar cell's surface. The plastic simply provides more light to the various solar cells.

Before cementing the top piece in place, connect flexible hookup wire to the remaining two wires of the solar cells. Make a slot or hole in one of the panel ends with the tip of the soldering iron for the two leads to extend through. Tie a knot in each wire so the wires will not be jerked out of the cells, causing damage and broken connections. Cut the wire ends to a 6 or 9 inch length (whatever length is needed) and solder small alligator clips. For 9 or 6 volt battery clips, you can pick up a set of plastic clips at most any radio electronics store.

Check Those Connections

When connecting plastic 6 or 9 volt clips to the solar panel, make sure the positive terminal is connected to the red wire.

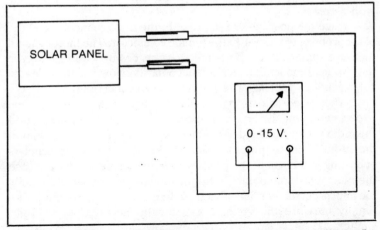

Fig. 17-6. Measure the voltage across the two connecting wires of the 14 cells. Under strong sunlight or a sunlamp you should have a reading close to 6 volts.

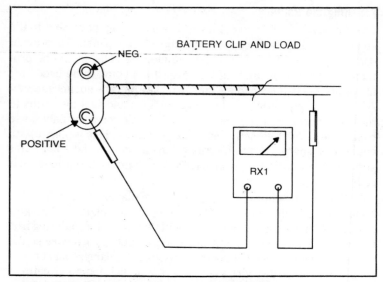

Fig. 17-7. Connect the red wire to the positive terminal and the black wire to the negative terminal of the solar panel. Check for correct polarity with a vom.

Generally, the negative terminal of the cell is connected to the black wire (Fig. 17-7). If in doubt, visually notice which wire goes to the small male connector. The red wire (positive) should go to the male connector and the black wire should connect to the female connector. Temporarily make the connections and measure the voltage with a vom.

Place the positive or red probe to the male connector and the black probe of the meter to the female connector. The meter should measure some voltage. If the meter hand goes backwards, reverse the two lead connections for correct polarity. Now you may clip the solar panel to your favorite radio battery connection.

When the radio is operated from the solar panel and you experience some distortion, suspect low voltage. This may occur if the radio is operated under a cloudy sky or under a reading lamp. If the radio is to be played inside, make sure the solar panel is operating in the sun or the panel is near an outside window. In case the radio does not operate in bright sunlight, either the radio pulls too much current or insufficient voltage is getting to the radio. Simply measure the voltage output with the vom under bright sunlight.

Here's to happy listening!

This project is simple to build and has only three working components. A red flasher LED is used to trigger the other two red LED lights. With two LEDs in series with a flashing LED, all three lights will blink. The supply voltage (6V) may be taken from one of the universal solar panel power supplies or you may build one on top of the small black box.

The flashing LED may be LED 1 or 3, it doesn't matter. Simply wire all three LEDs in series. Look closely at the bottom edge of the LED. You will find a flat or notched area. This area represents the cathode terminal (Fig. 18-1). The negative terminal of the solar cells will go to the cathode terminal with the positive connected to the anode terminal of LED 3. See Table 18-1.

When the solar power supply is connected to the LED string, the flashing LED will begin to flash on and off. The other two LEDs will follow suit. Choose LEDs with a typical forward voltage of 2.5 volts. The jumbo red and those with a fresnel lens, provide greater

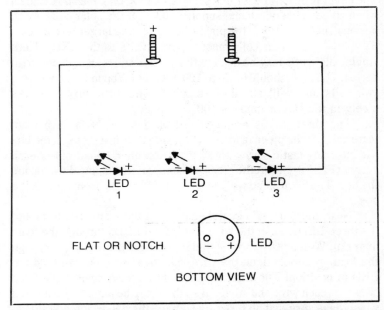

Fig. 18-1. Check for a notch or flat area at the bottom of the LED. The flashing red LED is number 1 with two other jumbo LEDs in series. When voltage is applied, all three LEDs begin to flash.

145

Table 18-1. Parts List for Project 18.

```
1 — Red flashing LED — Radio Shack #276-036
2 — Red jumbo LEDs — Radio Shack #276-041 or 276-070
                     Poly-Pak #92CV6541
                     GMI #18A18665
                     GC Electronics J4-940
                     ETCO #VA 353
Experimenter Plastic Box 4×2/4×2¼ Radio Shack #270-231
                     GC Electronics #H4-722
                     ETCO #158VA


Cost — (less Solar Panel) — $5.00
Construction time — 2½ hrs.
```

brightness. This black-blinking solar box has provided enjoyment to youngsters who stand watching the lights blink on and off.

Construction

Select a black plastic box with either a plastic or metal cover. The cover is discarded for this project. A 4" × 1½" × 2" box is ideal when an outside solar panel supply is used. If the solar cells are to be constructed on top of the plastic box, choose a larger box, at least 6" × 3¼". The solar cells must supply 6 volts to the LEDs. Here single solar cells may be used with a total of thirteen cells wired in series. The cells should be 50 to 100 mA units. You may choose six 1 volt cells that will take up less space. This little project works nicely in the sun or under a 100 watt bulb.

The three LEDs are mounted on a piece of ⅜ inch foam material. Cut the piece of foam to fit inside the plastic box. If the box has internal plastic screw holes, a section must be cut out of each corner (Fig. 18-2). Cut the piece of foam so it fits snugly down inside the box. The foam piece will end up ½ inch from the front edge of the box.

Now, measure off equal distances on the foam area to mount the three LEDs. Push the wire ends of the LEDs through the foam material. With a little care, both wire prongs will easily go through the foam material. Bend the wire ends over so they will hold the LED in position. The LED base should be flush against the foam area. Let each wire end stick out so they may be joined with hookup wire. Make sure that all three LEDs have the flat or notched side in one direction. The LEDs will not light if one or more are connected backwards. Mark the cathode terminal on the correct wire.

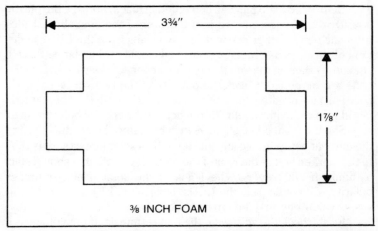

Fig. 18-2. The corners of the foam material may need to be cut out if the plastic mounting screw holes are inside the black plastic box. Cut the piece of foam so it will fit snugly inside the box.

Connecting the Components

Use bare hookup wire between the cells. Connect each cell in series with the next one (Fig. 18-3). Solder a three inch piece of wire to each remaining lead of cell 1 and 3. Connect these wires to the small bolts inside the black box. The ends of the ½ inch bolts stick out the back side of the box for connecting the alligator clips from the solar power panel.

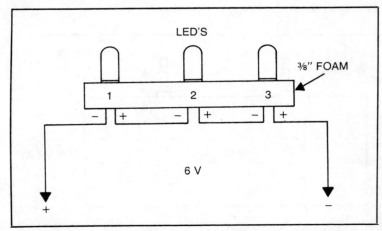

Fig. 18-3. Use bare hookup wire between the LEDs. Connect all three in series. Observe the correct polarity when connecting the LEDs together.

Before the foam piece is cemented in place, prepare the plastic box. Drill two ⅛ inch holes in the center of the box, about 1 inch from each end. Either connect the two leads from the LEDs under the bolt head or use metal eyelets to be soldered up later. Place a nut and washer on the small bolt. Now, connect the wire before the bolts are tightened. Check the polarity of the two wires and mark positive and negative beside each bolt (Fig. 18-4). The cathode marking of the foam panel will connect to the negative bolt terminal.

Slip the foam piece about ½ inch back into the plastic box. Try the unit out before going any further. Clip a solar power supply (6V) to the marked bolt at the rear of the box. All three lights should start to blink. If not, reverse the leads of the solar panel. Improper polarity will not damage the LEDs. In case the LEDs will not light, suspect that one may be wired backwards.

First, check the voltage at the bolt terminals. If the voltage is 6 volts or higher, check the polarity of each LED. You may quickly check the 1st and 2nd LED by shorting out the terminals on the 3rd LED. If the lights begin to blink, the 3rd LED is wired in backwards. Simply unsolder the connections, pull out the LED and reverse the terminals. Now, solder up the terminals and try again.

Always remember, the body of the flashing LED must be covered or it will not flash. Hold the body of the flashing LED between two fingers to shield the sun or light from the 100 watt bulb. This LED inside the black box will flash since light does not strike it. The front of the blinking black box will always be away from the light. After the lights are flashing, cement the foam piece ½ · inch from the front edge.

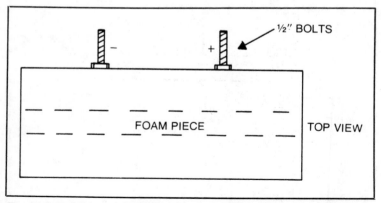

Fig. 18-4. Install two ½-inch bolts through the back holes. Mark positive and negative beside the correct bolt for easy hookup of the solar panel.

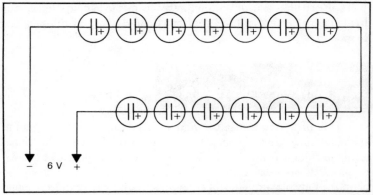

Fig. 18-5. All thirteen solar cells are connected in series for the 6 volt supply. Choose 50 to 100 mA cells for this project.

To make the flashing light of the LEDs appear brighter, a smoked or dark piece of plastic is placed over the front of the box. Salvage a piece of plastic from a cassette plastic case. Cut the piece of plastic the same size as the metal front cover. Mark where the four holes are to be drilled at the ends. Use a 3/16″ bit so the holes will fit over the plastic screw holes of the plastic box.

Connecting the Solar Cells

If by chance, you decide to place the solar cells on the plastic box instead of an outside solar panel, choose a larger box to mount all of the cells. You must connect thirteen single solar cells in series to acquire the correct operating voltage (6V). Choose 50 to 100 mA units. First, connect a ½ inch wire to the bottom of each cell. Place a dab of rubber silicone cement under each cell and press down in the plastic box. Line up all cells in the same manner. Let it set up overnight.

After the cells are set up, connect the bottom wire to the top hole of the next cell. You are now connecting all thirteen cells in series (Fig. 18-5). Connect a four inch wire to the top of cell 1 and the bottom of cell 13. These two longer wires will go through two separate ⅛ inch holes at each end of cell 1 and 13 to the corresponding LED connections. Measure the output voltage from the solar cells under a 100 watt bulb or in the sun. Check the cells for a reading lower than 4.5 volts. Suspect a cracked cell or poor soldered connection.

With the cells wired correctly and the front cover in place, set the black-blinking box in the sun or connect to a solar panel of 6 volts. Keep the face of the box away from the outside light. The lights begin to blink and will continue to do so until the blinking box is moved.

PROJECT 19
SOLAR EARPHONE SIGNAL TRACER

This little one-transistor signal tracer can trace the signal in any audio amp gear. If your CB receiver is weak, check the signal from the volume control to the speaker. The earphone tracer is ideal to check those weak audio stages. When one side of your stereo phono or tape player is low in volume, simply connect the earphone signal tracer and go from stage to stage. The solar signal tracer may be used as an audio amp for those crystal sets or one-transistor receivers. You may use this signal tracer to check out any audio problem in this project book.

The little signal tracer operates on from 1 to 3 volts of solar power (Fig. 19-1). Since the project pulls less than 1 mA of current, it is ideal for solar cell operations. In fact, to keep the price down, you may use up those broken cell pieces. The solar cells will even function under a bench light or reading lamp without the light being right down on the solar cells. See Table 19-1.

A pnp type transistor is used as the audio signal amp. The transistor circuit is protected by C1, a 600 volt blocking capacitor in case the signal tracer is used in tube circuits with higher voltages, the transistor will not be damaged. R1 may be used to control the volume in powerful audio output circuits. The small earphone is ideal to locate those weak audio stages; especially in locating an open or dried-up electrolytic coupling capacitor.

Signal Tracing

Let's take the case of a stereo receiver with a very weak right

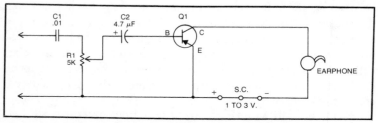

Fig. 19-1. A very simple one transistor circuit is found in this signal tracer.

Table 19-1. Parts List for Project 19.

C1 — .01 — 600 V capacitor
C2 — 4.7 vf 16V electrolytic capacitor
R1 — 5K volume control
Q1 — pnp low amp transistor, GE 53, SK3004, 2N3906 or equivalent
 2 — Banana jacks and plugs
Earphone — 2000 ohm impedance, earphone jack
S.C. — Solar cell 1 to 50 mA, 1 to 3V Solar Cell, GC Electronics
        #J4-801, Edmund Scientific Co. #42,710, H & R Inc. #TM 21K749.
        Solar Cell Chips, Poly-Pak #5310, Edmund Scientific Co.
        #P-30,828, and John Meshna Co. #H-13.

Cost under — $10.00
Construction time — 2 hrs.

channel (Fig. 19-2). Since the right channel is weak in radio, phono and tape operation, you may assume the trouble lies in the right audio circuit. Turn the balance control to the right channel and adjust the volume half way on. Now, start at the volume control of the right channel and signal trace through the right audio channel. Clip the common terminal to chassis ground and tune in a radio station.

You may note weak volume at the volume control. Open up the volume control if the signal is too weak and cannot be heard in the earphone. Go to the base of the first af amp stage (Fig. 19-3). Now, compare the volume on the collector terminal. The audio signal should be greater here. Go from base to collector of each amplifier stage. You may have to turn down the volume on the signal tracer for comfortable earphone reception.

Somewhere along the line, the volume will become very weak. Check the circuit here very carefully because you have located the weak stage. Weak audio stages may be caused by defective coupling capacitors. Simply check the signal on both sides of the capacitors. The signal should be the same on both terminals. Usually, a weak stage in the phono, tape player, or radio may be caused by an open input coupling capacitor.

Construction

Since there are very few components found in the solar earphone signal tracer, the unit may be constructed inside a small plastic mini-box. Any size will do if the box is big enough to house the small solar cells mounted in the topside. A four terminal lug was used to mount most of the critical components. Mount the volume

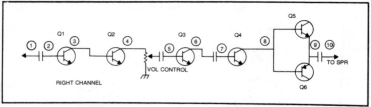

Fig. 19-2. Here is a typical stereo audio amplifier. There are ten test points. The weak stage may be compared at the same spot in the normal channel of the stereo receiver.

control and earphone jack on the front panel (Fig. 19-4). If the earphone is not to be used for any other project, the wires may be soldered directly into the circuit with the cord coming out of a hole in the front panel. The input alligator clip cable may be soldered directly into the circuit eliminating the banana jack and plugs, if so desired.

Prepare the front panel by drilling a ⅜ inch hole for the volume control. Drill two small ¼ inch holes for the banana jacks. Measure the diameter of the earphone jack and drill a hole to the left of the input jacks. If the front panel is metal and not plastic, the earphone and input jacks should be insulated. When the earphone and input wires are soldered directly into the circuit, without jacks, the metal holes should contain a rubber grommet to protect the wires from shorting against the front metal panel.

Temporarily place the solar cells on top of the mini-box. Leave the cells in their plastic containers for added protection. If broken solar cell pieces are used, place them in a plastic container. Drill

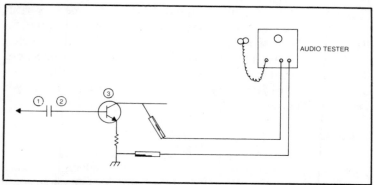

Fig. 19-3. Connect the alligator clip to the base terminal of the 1st audio amp and the common clip to ground. Now, go to the collector terminal of the same transistor, you should have a large gain in signal.

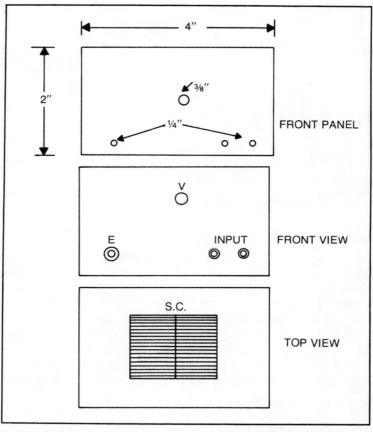

Fig. 19-4. Here are the front panel dimensions of the small signal tracer. All holes are drilled before any components are mounted. The solar cells are mounted on the top of the mini-box.

two holes through the plastic case and container. Feed the cell wires down through these holes and cement the plastic containers to the top of the plastic mini-box with rubber silicone cement. Connect the solar cells in series if separate cells are used (Fig. 19-5). Any size solar cells will work here. When chips or broken cells are used, select the larger pieces. Three or four separate pieces will do the job. Solder a piece of fine wire to the back of each cell. Connect this wire to the top bar of the next cell. All cell pieces may be connected together before they are set in a dab of clear silicone cement. Place them in a small plastic container to prevent breakage.

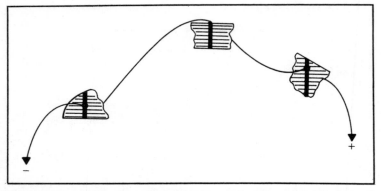

Fig. 19-5. Connect the solar cells in series if separate cells are used. Broken cell pieces may be used in this signal tracer project.

Wiring

First, mount all components on the four lug terminal strip (Fig. 19-6). Connect the various wires to the volume control and solar cells. Fasten the terminal strip to the back of the mini-box. Use silicone rubber cement to hold it in place or a bolt and nut through the back panel. Make sure the red or positive terminal of the solar cells go to the ground lug or emitter terminal. The negative lead connects to one side of the earphone.

Connect one side of C1 to the top of the volume control. The other terminal goes to the red banana jack or input wire. If the input

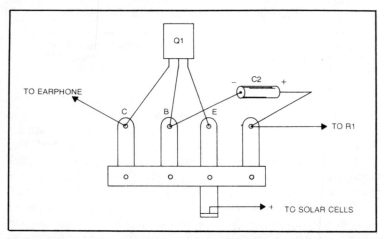

Fig. 19-6. Mount all components on the four lug terminal strip. Connect the various wires to the volume control and solar cells.

154

wires and the headphone cable is to be soldered directly into the circuit, tie knots in the cord so they cannot be pulled out. Now, solder a connecting wire from the collector terminal to the other side of the earphone.

Double check all wiring connections. You may check off all components and wiring as you go through the circuit in (Fig. 19-7). It's best to check all connections before firing up the project. You don't have to worry about turning off the switch or replacing batteries with this signal tracer. This little tracer will function under any type of light, even on cloudy days.

Testing

Simply connect the alligator clips to the ground and top of the volume control in the small portable radio. Turn the radio on and tune in a station. Now, turn the volume down and listen for the small audio signal in the earphone. Remember the signal is fairly weak at this point. Turn the volume up and disconnect the radio speaker. Now, go from transistor to transistor through the audio stages. As you approach the final amp at the speaker, the volume may be turned down on the radio or signal tracer. The audio signal is very loud at this stage if the radio is functioning correctly.

When no signal is found at the volume control, suspect improper wiring of components in the signal tracer. Check the voltage at the collector terminal of Q1. This voltage may range from 1 to 3 volts depending on the total number of solar cells in series. Suspect improper wiring of solar cells or cell breakage when there is no applied voltage. Don't forget to check for incorrect voltage polarity (−) at the collector terminal.

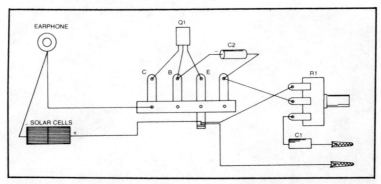

Fig. 19-7. This is a diagram of how the various components are connected. Cross off the components as you wire them into the circuit.

Go to the transistor circuits and check for correct voltage at the collector terminal. Double check the transistor connections. Visually, inspect the bottom terminal of the transistor. The collector terminal goes to the earphone with emitter terminal to common ground. C2 connects to the base terminal. With visual inspection and voltage measurements, any type of error is quickly located.

PROJECT 20
SOLAR AUDIO GENERATOR

This little audio signal generator produces audio signals at 1 kHz and 5 kHz. The 1 kHz signal is used for signal tracing and the 5 kHz is used for frequency response. The audio generator may be used to signal trace any audio circuit. Simply inject the signal at the audio amplifier and listen for the tone in the speaker. You may check for a weak or dead audio condition in a cassette or stereo-8 tape player, phonograph, radio, or audio amplifier (Fig. 20-1).

Only three regular solar cells or one 1.4 V solar cell powers the small audio generator. The circuit is built around a Radio-Shack LM3909 oscillator IC. Since the IC will work on very low voltage it is ideal for this solar project. Greater volume output may be obtained by adding another cell or two. This little test instrument does the job and is fun to build. See Table 20-1.

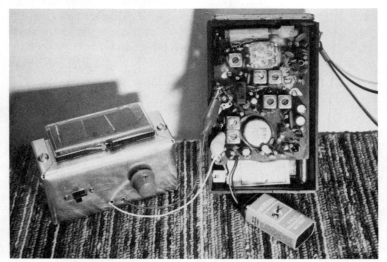

Fig. 20-1. Here the solar audio generator is checking the audio signal of a portable radio. Start injecting the audio signal at the volume control.

Table 20-1. Parts List for Project 20.

IC — LM3909 Radio Shack #276-1705
IC — Socket — Radio Shack #276-1995, GC Electronics #F2-1000
R1 — 5K volume control — Radio Shack #271-1720
         GC Electronics #B1-661
R2 — 68 ~ 1 watt carbon resistor
R3 — 680 ~ 1 watt carbon resistor
R4 — 100 ~ 1 watt carbon resistor
S1 — Sp ST toggle switch or slide switch
        GC Electronics #E2-117 (toggle)
        Radio Shack #275-615 (toggle)
        Radio Shack #275-407 (slide switch or equivalent)
        ET Co #0175W (slide)
T1 — 1K or 500K primary, Sec 8 miniature audio
  Output transformer — Radio Shack #273-1380
      GC Electronics #D1-712, GMI #D1-712
      ETCO #007XF
C1 — 50 μfd 16V electrolytic capacitor
C2 — 1 μfd 16V electrolytic capacitor
C3 — .005 μfd capacitor 200 V
Solar Cells — GC Electronics (1.4V) cells) #J4-803 or
      3 — .5 V cells (40 to 50 mA)
1 — Chassis box — 4×2⅛×1⅝ — Radio-Shack #270-239
         GC Electronics #H4-724
         ETCO #158VA
Perfboard — Radio Shack #276-1394
    GC Electronics #J4-612
    ETCO #0218C
2 — Alligator clips
Misc. — Solder, hookup wire, bolts and nuts, etc.

Construction Time — 6 hrs.
Cost — Under $20.00

Perfboard Construction

To cut the perfboard, layout the dimensions on one corner and mark with a pencil down through the row of holes (Fig. 20-2). Cut the piece out with a hacksaw or coping saw. If a vise is handy, clamp the perfboard where the saw cuts through the board. Be careful because sometimes the fiberboard breaks or cracks. Sand or use a pocketknife blade on the rough edges. All components except the solar cells, toggle switch, and volume control mount on the perfboard. The other components mount on the front metal panel.

Mounting the Components

Drill two ⅛ inch holes to mount the power transformer. Insert the prongs and bend them over. These are the only holes needed to be drilled in the perfboard since all other component leads tie through the small holes. Mount the 8 prong IC socket in the center of the board. Bend over two tabs on each end. Now, all other components may be mounted as they are wired into the circuit.

First, wire up all components on the perfboard (Fig. 20-3). Looking down on the IC socket, mark terminal 1 with a red dot from a marking pen. Turn the board over and mark the same contact (1) underneath with a red dot. This makes it much easier to connect all components to the IC socket.

Solder the blue wire of T1 to pin 5 of the IC socket. Ground the white wire of the T1 to the metal shield of T1. This area will serve as common ground. Cut off the center tap (CT) bare wire and tape up. Solder the green wire of T1 to pin 2 of the IC socket. The secondary red wire of T1 will go to the top side of the volume control after the perfboard is mounted in the metal cabinet.

Connect C1 to the positive terminal 5 and ground the common end to ground. Solder the positive terminal of C1 to pin 2 and the negative to pin 8 of IC socket. Push both R2 and R3 through the perfboard and solder to terminal 8 of the IC socket. Connect two

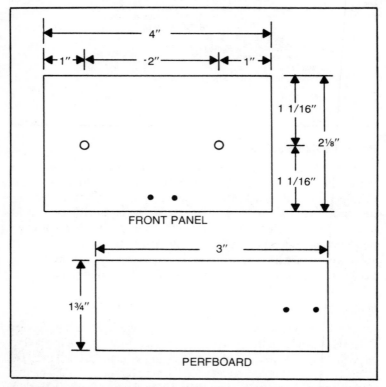

Fig. 20-2. Layout the perfboard dimensions and cut with the hacksaw. Be careful not to crack or break the perfboard.

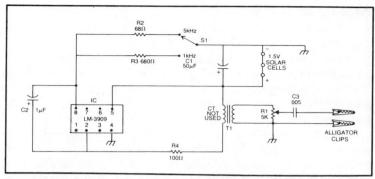

Fig. 20-3. The generator circuit is quite simple and uses only a few parts. Mark terminal 1 of the IC on both sides of the perfboard.

four inch wire leads to the other end of these resistors to go to the toggle switch in the front panel. Solder a ground wire to the shield of T1 and twist all three wires together. These wires will be soldered later to the toggle switch.

After the wiring has been completed on the perfboard, place a large dab of silicone rubber cement on the back of each corner. This will hold the perfboard in place and up from the metal box. Place the perfboard on the back and inside of the metal chassis box (Fig. 20-4). Insert a toothpick in the middle of each end so the board will not go down against the metal chassis while drying. Double check so no bare wires of the IC socket or other components are touching the metal side. Let the silicone rubber cement set up while preparing the front cover.

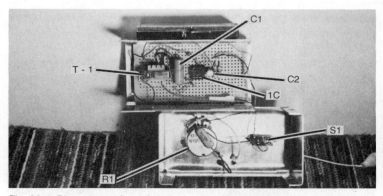

Fig. 20-4. Place the perfboard on the back and inside of the metal chassis box. Rubber feet hold the perfboard in place and away from the metal side.

Preparing the Front Cover

Stick a piece of masking tape down the center of the front cover. This will serve as a hole template and prevent scratching the front panel. Mark off the two ⅜ inch mounting holes 1 inch from each end. Center up the holes. First, drill a ⅛ inch hole and then enlarge it with a ⅜ inch bit. Drill two small ⅛ inch holes in the center bottom area for the two test leads. Ream out or clean off any metal burrs from each hole.

When a slide switch is used instead of a toggle switch, the hole area will have to be filed out with a rectangular rat-tail file. Clear out an area ½ × ⅜ inches for the slide switch to work in. Place the switch on the front surface and mark the two side mounting holes. Drill two ⅛ inch holes to mount the slide switch. Now, mount the switch and volume control on the front panel.

Connecting the Solar Cells

You may use three separate 40 or 50 mA solar cells or one 1.5 volt cell. The solar cells should be placed in a small plastic box. Mount the 1.5 volt cell in the plastic box it comes in. Drill two holes in the end of the plastic box. Set the box on the metal top chassis and mark the holes. Drill two ⅛ inch holes through the metal (Fig. 20-5).

Connect the three cells in series before mounting in the plastic box (Fig. 20-6). Insert the red and black wire through the plastic holes and through the metal chassis holes. Close the lid on the cells. Apply a dab of silicone cement on the bottom of the solar cell box to hold it to the metal cabinet. Solder the black wire (−) of the top of the solar cell to common ground. Connect the red wire (+) of the bottom of cell three to pin 5 or the positive terminal of C1. The wiring of the solar cells should be done before the plastic box is cemented to the metal chassis.

Fig. 20-5. The solar cells are mounted inside the plastic box in which they came. Cement the plastic box to the top with rubber silicone cement.

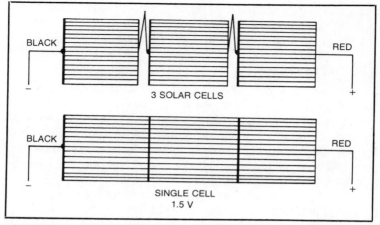

Fig. 20-6. If single solar cells are used, connect them in series. The red wire connects to the black wire of the next cell. If the cells do not have connecting wires, solder a black wire to the top and red wire to the bottom.

Solder the black or common ground lead from the perfboard to the common terminal of the toggle switch. Connect the other two leads to the remaining terminal. If a slide switch is used, solder the black (ground) lead to the center terminal and the other two leads on the opposite ends. It doesn't matter which wire goes to either outside terminal. When switched in the highest tone will be marked 5 kHz.

Now, run another wire from the black or common terminal of the switch (S1) to the low side of the volume control. Connect the red wire from T1 to the top of the volume control. Solder C3 to the center terminal of R1 and the alligator clip wire. C3 may be mounted with a dab of rubber cement to the front panel. Connect the ground lead of one alligator test lead to the bottom side of R1. Tie a knot in each test lead so the cord will not pull out.

Testing

Before placing the two metal box pieces together, check the unit out. Turn the backside upright so a 100 watt bulb can be placed over the solar cells. The light may be only a foot away to make the generator function. Connect the green lead to the center terminal of a volume control of a small radio. Ground the white alligator clip.

You should hear a loud sound in the speaker. Check both frequencies by sliding S1 or flipping the toggle switch. The highest tone is the 5 kHz and the lower the 1 kHz signal. This audio signal may be used to signal trace the audio circuits in radios, amplifiers, tape, or phono players. Just connect the white terminal to ground and the green lead to the audio input terminals.

No Operation

In case the generator does not function, check the voltage at pin 5 of the IC. You should measure 1.39 volts with a 100 watt bulb. Lower or no voltage at pin 5 may indicate a leakage or poor cell connection. Remove the red wire from the solar cells and measure this voltage. A lower than normal voltage indicates a cracked cell or poor cell wire connections. Visually inspect the cells and connections.

If over 1.35 volts is found on pin 5 and no tone, check pin 4 for proper ground connection. Measure the voltage on pin 4. If any voltage is measured here, pin 4 has a poor ground connection. Make sure R4 (100 Ω) is wired in the circuit). The generator won't oscillate with the resistor out of the circuit. Go over all connections. Check off each connection shown in the Fig. 20-7.

You will note the tone of the generator will change somewhat when the light is closer or pulled away from the solar cells. A louder tone may be heard by adding another solar cell. R1 may be eliminated if the output volume is not controlled. The value of R1 may be 500 to 1 k. A 5 k volume control was used here since it was found in the junk box.

The small audio generator may be used to check the sound stages of any type of radio. Start at the volume control and inject the audio signal. Go to the base and collector stages of each audio transistor until you hear a tone in the speaker. Now, back up a step. For instance, if you have the tone on the collector terminal and no signal on the base of the 2nd af transistor, suspect a defective transistor. When a signal is heard on one side of a coupling capacitor and not the other, suspect an open capacitor.

Suspect a defective tape head or audio amplifier section when one side of the 8-track tape player is weak or dead. Connect the audio signal generator to both sides of the tape head to determine which channel is defective. Leave the volume control at least half way open. After locating the defective channel, you now can signal trace each state with the small signal generator. Just go from base to collector terminal of each transistor stage. Use the 1 kHz tone for signal tracing purposes.

When the recording or playback is weak in a cassette player, suspect a dirty tape head of amplifier section. Clean up the tape head

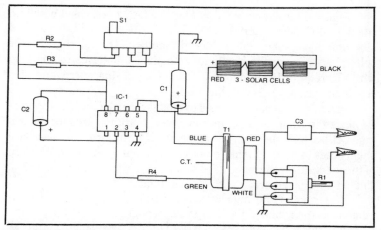

Fig. 20-7. Here is a pictorial diagram showing how to connect all the components. Check off each connection as it is soldered into place.

with alcohol and a cleaning stick. Now try the cassette once again. If the signal is still weak, connect the signal generator leads to the back of the tape head. If it is still the same, go from the coupling capacitor to the base of each transistor stage. The audio signal should be the same on both sides of a coupling capacitor. If not, replace the capacitor.

You can quickly check out that suspected phono cartridge with this solar generator. If one phono channel is weak or dead, suspect a defective cartridge or amplifier section. Connect the generator to the crystal cartridge terminals. Check both sides. The volume should be the same. Replace the crystal cartridge when the volume is the same. Signal trace the amplifier section when the volume is lower in one channel. Use the 5 kHz tone to check out the frequency response of the amplifier. Any vibration of the speaker or components at 5 kHz indicates a torn or loose speaker cone.

PROJECT 21
SOLAR POWERED CALCULATOR

You can purchase a small solar powered calculator for about $20. How about solarizing that small calculator you now own? Since these units pull only a few milliamps of current, small chips or pieces of solar cells may be used to power them. Just connect them in series to get the required operating voltage.

Figure 21-1 shows a Sharp Model EL-211 solarized with one small 3 volt microgenerator cell. This small solar unit consists of nine individual solar cells soldered together. The microgenerator cell or other connecting cells may be placed in series to come up with the required voltage. The actual cost of this microgenerator cell is about $4.50. See Table 21-1.

Remove the small batteries to mount the solar unit. Temporarily place the solar cells on the top cover. Mark the area for mounting. Drill or use the point of a pencil soldering iron to make two small holes underneath the cells. A small needle heated with the tip of the soldering iron can be used to make the wire connecting holes. Cut off any raised areas with a pocketknife.

Now, mount the microgenerator cell. Place a layer of clear silicone cement under the cell area. Straighten up and press the cells into place. Do not press too hard or you may damage or break the cells. Simply press the cell into the cement with a pencil point or knife blade.

After the cell has set up overnight, wire the solar cell to the battery terminals. If the small connecting wires are not long enough, extend them with a very small flexible hookup wire. Use spaghetti insulation over the bare wires and solder all connections. The wires may be soldered directly to the battery terminals.

Remember the top wire of the solar cell is always negative and the bottom connecting wire is positive. If in doubt, use the vom to measure the output voltage with the red lead from the meter as the positive terminal. Simply reverse the two leads if the meter reads backwards. Now, the lead from the solar cell touching the red lead of the vom is positive.

Fig. 21-1. Here is a Sharp model EL-211 calculator powered by a 3 volt microgenerator solar cell. Mount the cell directly upon the front.

Table 21-1. Parts List for Project 21.

Construction Time — 3 hrs.

Material — Edmund Scientific — 1-3V microgenerator
GC Electronics — 3 — J4-801 — 1 volt cells or
2 — J4-804 — 1.4 volt cells
Chips and pieces
Edmund Scientific — 50 pieces No. P-30828 — $9.95
25 pieces No. P-42,749 — $9.95
or small broken pieces from solar cells.
Flexible Plastic — Coffee, Crisco, Oleo cover — cut to size 2″ × ¾″.

Take a peek at the battery connections to determine where the positive wire goes. Most battery polarity terminals are marked in the battery area (Fig. 21-2). Usually, the negative terminal has a spring. Check and see if a black wire is connected to the calculator. The metal indented terminal is the positive lead. A red wire may connect from this lead to the circuits of the calculator. Solder the positive solar cell lead to the positive battery terminal and the negative lead to the negative terminal.

The solar powered calculator operates beautifully in direct sunlight. In fact, this unit operates nicely even on cloudy days. You can operate the calculator under an incandescent lamp with good results. Of course, the calculator is not as effective under a fluorescent shop light.

Constructing Your Own Supply

To determine how many cells are to be used, you must know the operating voltage and current of the small calculator. Add the total battery voltages. Insert the vom in series with the batteries to

Fig. 21-2. Take a peek at the battery connections. Solder the positive lead to the positive battery terminal. Do likewise with the negative terminal.

measure the correct current (Fig. 21-3). This small Sharp calculator operated at 3 volts with 40 microamps of current. These small calculators pull very little current and almost any solar cells will operate the unit nicely. In fact, pieces and sections of solar cells were used to make up the small solar power unit. These pieces can be purchased separately or you may use broken pieces of left over solar cells. Originally these cell chips were intended for solar watches and other devices. Usually, the current of these cells may be from 1 to 6 milliamps.

Select solar pieces with a large connecting strip on the top side of the cell (Fig. 21-4). They are easiest to connect and wire up. You may use any pieces with strips, except all of the strips must be connected together. Scrape the cell strip area with the blade of a pocketknife. The scraped area should look very bright. Place a small bead of solder on the area with the point of a small soldering iron. Touch the spot with the iron and then feed solder to the point. Only a small bead is necessary to connect the strand of wire. If too much solder or a high point of solder is found, the cells may not mount evenly. To make a certain cell fit in a given area, chip away on the ragged edges with scissors or a side cutter. If the cell breaks down the middle, start over on another one. You will find these cells are very brittle and break easily.

Since this small calculator operates at 3 volts, nine small pieces of solar cells were used. It's best to have at least an extra 1

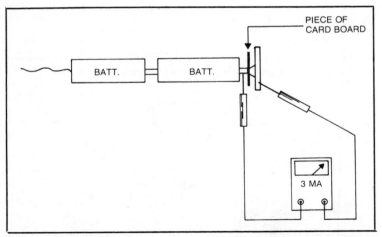

Fig. 21-3. To check the current of the small calculator, place a milliampere meter in series with one of the batteries. Use the 3 mA scale since most of these units pull very little current.

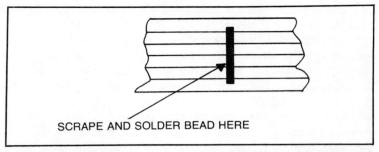

SCRAPE AND SOLDER BEAD HERE

Fig. 21-4. Select solar chips or pieces with a large connecting strip. They solder and wire up a lot easier than the thin-lined ones.

volt or so besides the needed operating voltage when used inside the house. Actually, the nine cells should produce around 4.05 volts. These nine solar chips were wired in series as shown in Fig. 21-5.

If a calculator operated at 4.5 volts with 1 mA of current, you would need twelve solar pieces connected in series. Don't worry about the current as most solar chips have a greater current carrying capacity than needed for a small calculator. The biggest concern is to mount the needed pieces in a certain area on the case of the calculator. Make sure the cells can be on the front area.

Connecting the Cells

After selecting and choosing how the solar cells will line up in a small space ¾ inch × 2 inches, start connecting the cells together. Begin with the top of one cell and solder this wire to the bottom of the next cell. Try to keep the cells uniform and in a straight line. If the cells are too close to keep in line, unsolder and pull the cell away. If not, they may pile upon one another.

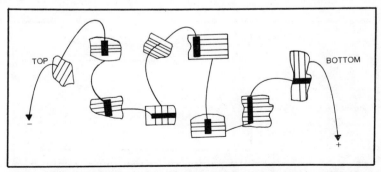

Fig. 21-5. Here the nine solar chips are wired in series. Connect the top of one to the bottom of the next cell. Always make a good clean solder connection.

167

Use the knife blade to hold the small connecting wire on the correct spot while soldering. Lift the small wire on top of the next cell with the knife blade. Snip off the small strand of wire with scissors or side cutters so they will not short out against the next cell. You may have to add another bead to the top when the bottom wire is connected. Sometimes excessive heat will melt down this connection.

The cells are easily connected and soldered together if placed on a small piece of cardboard or block of wood that can be rotated. Use a pocketknife to turn the cells over to make the other wire connections. Now, rotate the cardboard or wood piece to make for easy soldering of the connections.

Mounting the Cells

The solar cells may be mounted directly on the plastic case or if a metal decal has to be covered, use a separate piece of plastic for mounting. Cut a piece of flexible plastic from a coffee can plastic lid with a pair of scissors. Here a 2 × ¾ inch strip was cut out to mount the nine solar pieces (Fig. 21-6). Place a layer of clear silicone cement on the plastic piece. Spread it out with a pocketknife. Too much cement will ooze out and cover up the cells. Lay the cells down in order on the cement area. Press each cell into the cement with a pencil or knife blade. Be careful not to damage or break any of the cells. With the pocketknife blade, clean off any excess cement. Let the solar unit set up overnight.

Now solder two long leads to connect the cells to the battery terminals. Measure the voltage and it should be over 3 volts under a reading lamp. Under direct sunshine, the voltage should be around 4 volts. If not, go over each cell and check for poorly soldered connections.

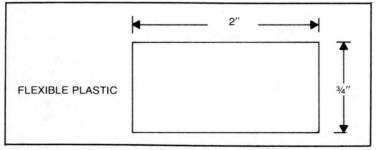

Fig. 21-6. Place the nine solar chips on a 2 × ¾ inch plastic strip. Cut the piece of flexible plastic from the top of a coffee can cover.

Fig. 21-7. Here is the finished project. This calculator works beautifully in the sun or under an incandescent lamp.

Drill or make two small holes where the solar power unit will mount. Keep the end hole underneath the length of the cells, so they will not show after cemented into position. Place three separate areas of plastic cement on the rear of the plastic solar piece. Now, feed the two wires into the respective holes and press the solar unit in place.

Wipe off any excess silicone cement. This cement is like rubber when it sets up and must have a smoothed finish appearance. Bevel the edges if necessary (Fig. 21-7). One nice thing about this calculator, you don't have to ever worry about turning it off. There are no batteries to wear down or lead out. In fact, this calculator will operate for years to come without any costly service.

PROJECT 22
CHILDREN'S SOLAR BLOCKS

Most little children respond to flashing lights and unexpected noise. When the favorite TV commercial is flashed on the screen, all children stop playing and watch those commercials because of the increase in the volume of the sound. Since little children like to play with blocks, three large plastic blocks were constructed with blinking lights and unusual sounds. Just set the youngster on the living room carpet where the sun pours in from a large window and let them build with these electronic blocks.

The small plastic blocks were constructed of plastic picture cubes. These plastic cubes are used to house four different pictures mounted inside the plastic area. They are the ideal size with heavy plastic construction. Any type of plastic boxes may be used, as long

169

as they resemble a set of blocks and can be stacked. Check the various variety and department stores for these photo cube boxes.

The Red Flasher

Here is a very simple circuit utilizing a red LED flashing diode and solar cells (Fig. 22-1). When the sun strikes the solar cells, the small red LED starts to flash. Cut out a photo of a dog or cat and place the red LED as one eye. Cement the picture inside the plastic cube area. Hold the LED in place with black silicone cement around the LED and plastic picture frame area. Mount the solar cells in the top side. The other sides of the plastic cube contain other pictures or may be painted with bright colors. See Table 22-1.

Mount the 9 solar cells on a thin piece of foam material. You may also use four 1 volt cells. Cut the foam piece so it will fit inside the cube area. All solar cells are wired in series for the 3 to 5 Vdc operating voltage of the red flashing LED. When the solar cells are exposed to the sun, the LED will flash. The LED should be mounted to one of the side panels, so when the solar cells are upright or sideways to the sun, the LED can be seen.

Cut a piece of ⅜ inch foam material to fit in the top side of the photo cube. Now, lay the nine 100 mA or higher current solar cells in line (Fig. 22-2). Solder a tie wire to the bottom side of each cell. Mount three cells in a row. Apply a dab of cement to the back side of each cell. Bend the tie wires upward and mount each cell in line. Let the cells set up overnight.

Connect a four inch piece of flexible hookup wire to the top side of cell number 1 and the bottom side of cell number 9. These two

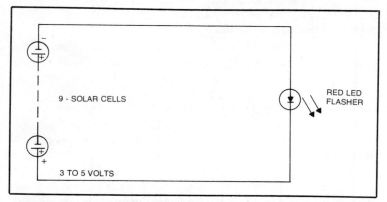

Fig. 22-1. Here is a simple circuit utilizing a red flashing LED and solar cells. The LED begins to flash when the sun strikes the cells.

Table 22-1. Parts List for Project 22.

The Red Flasher
        9 — solar cells (50 or 100 mA) Poly-Paks #5306
           John Meshna Co. # H-11, H & R Inc. #TM21K666,
           Solar amp #S122, GC Electronics #J4800 or J4801
        1 — Flashing LED — Radio Shack #276-056
        1 — Plastic photo cube — found at photo and department stores.
        Misc. — Wire, solder and silicone rubber clear cement
        Construction time — 4 hrs. — Cost — $22.50

The Whistler
        Solar cells (50 or 100 mA) Poly-Paks #5306
        John Meshna Co. #H-11, H & R Inc. #TM21K666,
        Solar Amp #S122, GC Electronics #J4-800 or J4-801
        Edmund Scientific Co. #42,268
        1 — Sonalert — GMI #12A2841-2 (1-4V c)
        Plastic cube — located at photo and department stores.
        Misc. — Wire, solder and silicone rubber cement
        Construction time — 4 hrs. — Cost — $25.00 to $30.00.

The Buzzing Bee
        4 or 5 Solar Cells — (25 to 100 mA) Edmund Scientific Co. #30,735
        Poly-Paks #5306, John Meshna #H-11, H & R Inc. #TM 21K666,
Solar amp #S122, GC Electronics #J4-800 or J4-801 (J4-801 1 volt type)
        1 — IC LM 3909 — Radio Shack #276-1731
        1 — Plastic cube
        1 — 8 pin IC socket Radio Shack #276-1995
        1 — 8 ohm pm Spk — 1½ to 3 inches (any size).
        C1 — 50 vfd 16V Electrolytic Capacitor
        C2 — 1 vfd 16V Electrolytic Capacitor
        R1 — 6.8K ohm fixer resistor — ½ watt
        R2 — 100r Fixer resistor ½ watt

        Construction time — 6 hrs.
        Cost — $25.00

Cost of each cube project will depend on what components may be solarized from the junk box.

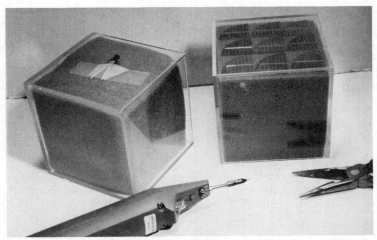

Fig. 22-2. Mount the nine solar cells on a piece of thin foam and place it in the top of the plastic cube. Dab each corner with silicone rubber cement to help prevent damage to the solar cells.

171

longer wires will solder directly to the LED terminals. The bottom side wire will connect to the positive side of the LED. Visually observe the LED for correct polarity (Fig. 22-3).

Check out the flashing LED before the cells are placed inside the cube area. Place a 100 watt bulb above the solar cells, the LED should start to flash. If not, reverse the two wires to the LED. Now, check once again. It's best to apply black silicone cement around the base of the flashing LED so outside sunlight will not strike against the lower body area. Sometimes, the LED will not flash if bright light is shining on it.

After the LED and solar cells are functioning, apply a dab of clear silicone cement to each corner of the solar panel. This will keep the cells from touching the plastic end which takes the hard knocks and prevents possible cell breakage. Some of these plastic cubes have a piece of foam rubber to hold the picture against the plastic cube. Replace the rubber foam material and cement a plastic end to the plastic cube. Now the cube is ready to flash.

The Whistler

The whistler lets out a 2.7 kHz tone when the sunlight strikes the solar panel. No switches are found here. Only two separate components, solar cells and a Sonalert® device are used to create a noise. The circuit is similar to the Sun-up Alarm.

Again, the nine solar cells are placed upon a ⅜ inch piece of foam material. The foam is cut to fit inside the top panel, so the solar cells will be protected behind the plastic. This solar panel is constructed in the very same manner as the Red Flasher. Instead of using nine, 100 mA or greater cells, select four 1 volt solar cells if desired (Fig. 22-4).

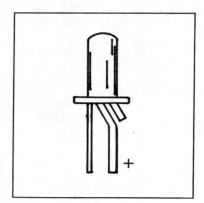

Fig. 22-3. Check the LED for correct polarity. When the cell doesn't begin to flash, reverse the two leads. Keep the body of the flashing LED out of the direct sunlight.

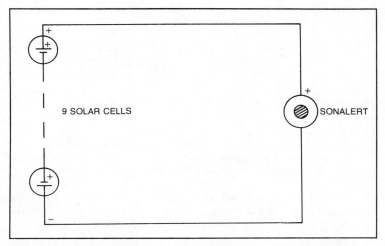

Fig. 22-4. The whistler circuit is somewhat like the Sun-Up Alarm. Mount the small Sonalert sounding device inside with rubber cement.

The two wires from the solar panel will connect directly to the Sonalert®. Place the bottom lead wire of the solar cell (+) to the positive connection of the sounding device. It's best to connect the Sonalert® and check it out under a 100 watt bulb before cementing into position.

Drill several ⅛ inch holes to fill the area of a quarter coin. Place the quarter on one side of the plastic cube and mark it. Center the quarter within the side panel. Another method is to place two rows of masking tape down the middle plastic side and lay the quarter right in the center. Now, drill about 10 holes in this area. The masking tape will prevent scratching of the plastic cube.

Dab clear rubber silicone cement on each corner of the foam panel before mounting in the topside. The solar panel must be in place before the Sonalert® device is installed. Now, place the sounding device over the drilled holes inside the plastic cubes. Place clear silicone rubber cement around the outside edge. Several layers of cement will hold the Sonalert® in place. Let the cement set up overnight before inserting the rubber foam material. The whistler is now ready to cut loose.

The Buzzing Bee

Here is a solar block circuit that sounds like a buzzing bee or insect. The tone of the oscillator changes depending upon how much light strikes the solar cells (Fig. 22-5). Actually, the 8 pin IC will

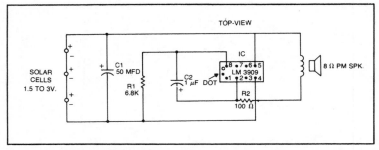

Fig. 22-5. The tone of the oscillator depends upon how much light strikes the solar cells. The LM3909 IC operates on 1.5 to 3 volts.

function with a very low voltage operation (1.5 to 3 V). The current drain is less than 0.5 mA so it's ideal for solar operation. Four 25 mA solar cells will operate the buzzing bee very nicely.

There are only two capacitors, two resistors, a three inch speaker with an 8 pin IC (LM3909) to produce the insect tone (Fig. 22-6). The size of the speaker may be any 8 ohm type. A small 8 pin socket should be used to mount the various components. Mount the IC when all components are wired and double checked in the circuit.

Select four 25 mA or higher solar cells. All four cells are wired in series (Fig. 22-7). Greater volume may be obtained by adding one or two more solar cells. The cells should be mounted on a ⅜ inch piece of foam and cemented to the top panel as in the other cube experiments. Leave two six inch leads to connect to the IC circuit.

Preparing the Speaker Area

Before mounting the solar cells, place masking tape to the middle side panel of the plastic cube. Choose any side at right angles with the solar cells. Lay the speaker on the tape and draw around the

Fig. 22-6. Only a few components are needed for an audible tone from the Buzzing Bee. Choose any size 8 ohm speaker that is handy and will fit inside the plastic cube.

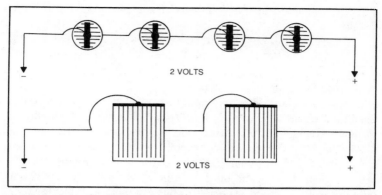

Fig. 22-7. All four solar cells are connected in series. You may use two or three 1 volt cells instead.

speaker frame. Drill ⅛ inch holes inside the speaker circle. By drilling many small holes, sharp pointed objects may not be pushed into the speaker cone. Try to space the holes ¼ inch apart.

Now, remove the masking tape. The tape helps to remove small pieces of plastic and prevent scratching the plastic cube. Be careful when drilling, so not to break or crack the plastic case. Place the speaker inside and see how the holes line up. The speaker may be cemented with silicone rubber cement inside the plastic panel.

After all wiring has been completed, check the circuit over at least twice. Now, connect the small speaker. Connect and test the IC oscillator before mounting inside the plastic cube. Place a 100 watt bulb over the solar cells and listen for a tone in the speaker.

If the speaker is dead, check the polarity of the solar cells. Measure the voltage at pins 4 and 5 of the IC socket. The voltage should be from 1.5 to 2 volts. Check each solar cell for possible cracked areas. When six solar cells are used, you should measure 2 to 3 volts at these connections. Check the connection of C2. This capacitor provides more volume and a richer tone. You will notice when the light is close to the cells, the tone is higher and with the cells farther away from the light, the tone is lower.

With the little buzzing bee circuit operating, cement the solar cells to the top of the plastic cube. Place a dab of silicone clear rubber cement to each corner. Mount the small speaker by placing a layer of clear cement around the outside of the speaker. If a 2 or 3 inch pin speaker is used in this experiment, silicone rubber cement does a nice job. Now place some cement on the back of the speaker and place the IC socket and capacitor on it. Let the speaker and components set up overnight. The base connection of the socket will be embedded in the cement and not short out against the speaker.

Check the unit out with a 100 watt bulb above the solar cells. Place a piece of plastic over the bottom side and cement to the plastic cube. Use either plastic cement or clear rubber silicone cement to bond the cube together.

Several solar cubes may be constructed with any one of the described circuits. A total of four or six plastic cubes should be made for stacking. The plastic blocks may be large plastic building blocks or any type plastic cubes that will contain the required components. Keep the outside areas of the blocks smooth with no bolts or nuts through the plastic. Building the solar blocks is not only fun and a learning experience, but watching a youngster play with them will be a joyful occasion.

PROJECT 23
A BLINKING-BEEPING BOX

Here is a novel solar project that sounds off with a beeping sound and flashing light. A flashing LED is used in the solar power circuit to interrupt the sound. In turn, the LED begins to flash producing a blinking-beeping sound (Fig. 23-1). The IC oscillator (LM3909) drives a small 8 inch pm speaker. When set in the sunlight or under a 100 watt bulb, the black box beeps and flashes until the light is taken away. See Table 23-1.

Again in this project you have no batteries to worry about or to shut off. The black box operates from solar power. The project operates on 5.5 volts and pulls less than 5 mA, making the beeping-flashing box inexpensive to build because you can use broken pieces of solar cells.

Fig. 23-1. Here is the front view of the beeping-blinking black box. Only the speaker and flashing LED appear on the front panel.

Table 23-1. Parts List for Project 23.

IC — LM 3909 — Radio Shack #276-1705
Flashing LED — Radio Shack #276-036 or equivalent
IC 8 pin socket — Radio Shack #276-1995
                    GC Electronics #F2-1000
8 ohm pm Speaker — Any size
C1 — 4.7 vf 15V
C2 — 50 ufd 16V
R1 — 370 Ω ½ watt fuse resistor
R2 — 100 Ω
Black Box — Radio Shack #270-233, ETCO #158VA
             GC Electronics #4-722
Solar Cells — 12 broken cells (5 mA or greater)
             Edmund Scientific #P-42,749
             John Meshna, Inc. #H-13, and Poly-Pak #5310
Misc. — Perfboard, hookup wire, solder, bolts & nuts, etc.
Cost — Under $15.00
Construction time — 4 to 6 hrs.

Selecting the Cells

Since the blinking-beeping box pulls less than 5 mA of current, this little box may be powered with solar chips or pieces. In fact, you may purchase a package of cracked or broken cells for less than two dollars or salvage what you may have already broken. These kits usually include 10 pieces, but you can break some of the larger ones, to make twelve separate solar cells.

Cut twelve 2 inch pieces of connecting wire. Use bare or covered wire. Radio Shack's wrapping wire is ideal for this project. Use the iron to melt off the insulation at one end. Hold the wire down upon a piece of cardboard or magazine and apply the iron to one end. Pull on the wire and the insulation comes right off. Now, solder one lead to the back of each cell except cell number 13. Here you will connect a six inch length of wire. Solder the wire to the center edge of each cell.

After all wires are soldered to the bottom side, arrange them on the back of the plastic box. Keep them in line and close together. Draw a rectangular area 2½ × 4 inches and keep the cells within this area. Make sure the rectangular lines are centered at the back of the box. Place a dab of clear silicone rubber cement on the back of each cell and stick to the plastic surface. Leave the connecting wires straight upward. Keep each wire to the right side of each cell of the row. The rest will follow in line. Let the cells setup overnight.

In case one of the broken cells may have a crack in the top area, make sure the bottom soldered area is together. To be sure, place a drop of solder over the cracked area. Go to the top of the cell and tie the small lines together with a fine wire. You may find a cell cracked on top, but not broken into. Top side wire repair may salvage these broken or cracked pieces.

Connecting the Cells

All twelve cells are connected in series (Fig. 23-2). Start at the top left-hand corner with cell number 1. Pull the connecting wire over to the center bar of cell number 2. Cut off with side cutters. Melt off the insulation tip with the soldering iron. Now lay the wire flat and solder to cell number 3. The wire may be held in place with a pencil or toothpick. Connect each wire in the very same method. When only lines are found and no bar area, solder the wire across all of the thin lines. Before soldering the wire to the thin lines on the top, apply solder on each line and leave a solder bead. Then lay the bare wires across these beads and solder up. After all connecting wires are in place, check for cracked cell areas. If one is found, bridge the thin lines with bare wire and tie to the connecting wire.

Solder a six inch piece of wire to the top of cell 1. Likewise, connect a six inch piece of wire to cell 12. Push these wires down through a 1/16" hole beside cell 1 and 12. If the holes are not drilled before the cells are placed on the plastic area, be very careful not to damage the cells. The top wire is the negative and the bottom wire in the positive terminal.

Before enclosing the cells for protection, check the voltage output with the vom. Set the meter to 15 volts dc. Place a 100 watt

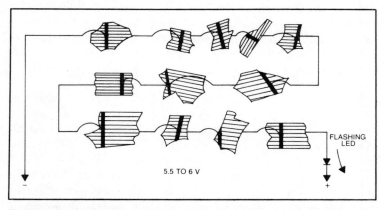

Fig. 23-2. Connect all broken cells in series. Start at the left-hand top corner with cell number 1. Connect the wire from the underneath side to the top bar of cell number 2, until all cells are connected.

Fig. 23-3. Since the cells are on the back side of the black box, they should be covered for protection. Cut a piece of clear plastic to cover the cells. Seal with clear or black silicone rubber cement.

bulb above the solar cells. You should measure 5 to 6.5 volts without any problems. If the voltage is lower than 5 volts, suspect a poor wiring connection or cracked cell. Observe which wire is positive when taking the voltage measurement. Mark this wire with tape for easy reference.

The cells should be covered so they are not easily broken. Cut a piece of cardboard at least ¼ inch thick and ¼ inch wide. One quarter inch rubber foam or foam material will do. A rectangular piece 4½ × 3¼ inches wide will do nicely (Fig. 23-3). This will enclose the entire solar cell area. Cement this strip to the back area. Now cut a piece of clear or indented plastic the same dimensions. This plastic piece will fit over the entire solar cell area. Cement this plastic piece to the cardboard. Let the rubber cement setup for a few hours and trim the edges with clear or black silicone rubber cement. Wipe off excess rubber cement with a paper towel. Not only does it have a neat appearance, but the dust and moisture is kept out of the solar cell area.

The indented plastic piece may be cut with a hacksaw, but be very careful not to crack or break the plastic. Turn the plastic smooth surface up. Draw the 4½ × 3 outline on the plastic with a ball point pen. You may want to try another method of cutting. Take a straight metal edge and lay it on the pencil line. Use the hot soldering iron tip and go along the metal edge. You may have to go over the same area several times. Lay the line over the edge of a table or desk and the plastic will break right off. Trim up the edges with a sander, if the edges are irregular. Otherwise, trim up the edges with a pocketknife.

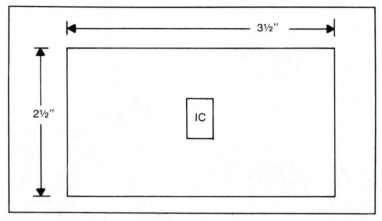

Fig. 23-4. Cut a piece of perfboard 3½ × 2½ inches. The perfboard may be cut from larger stock with a hacksaw or pocketknife. All parts are mounted on the perfboard except speaker, solar cells and flashing LED.

Preparing the Chassis

Cut a piece of perfboard 3½ × 2½ inches. These boards may be cut with a pocketknife or hacksaw (Fig. 23-4). When small pieces are used, mark the dimensions off on the perfboard. Lay a metal edger down the line of holes. Use a sharp pocketknife blade and go down across the holes. Break the piece of perfboard across the edge of a table. The narrow pieces are easier to cut off and break than big pieces. When cutting the piece from a large perfboard, use the hacksaw for the lon cut. The narrow piece may be cut with the pocketknife method.

Mount the IC socket in the center of the perfboard. Bend over a metal tab at each end. Mark terminal 1 on both sides of the board. This makes for easy terminal connections in wiring up the IC circuit (Fig. 23-5). The large capacitor may be mounted by bending over the wires before soldering it into the circuit. All small components are mounted as they are soldered into the circuit (Fig. 23-6). These board components may be connected in a very short time. Solder two six inch leads in the speaker and wire in series with R2(100 Ω) and terminal 5 of the IC. Run a four inch wire from pin 5 or positive side of C2 to tie to the side of the flashing LED. Go over the wiring at least twice before connecting the solar cells.

Check the unit before placing the perfboard into the plastic cabinet. Temporarily connect the negative terminal of the solar cells to the perfboard common ground terminal. Connect the LED in the circuit. Make sure the flat side of the LED connects to the

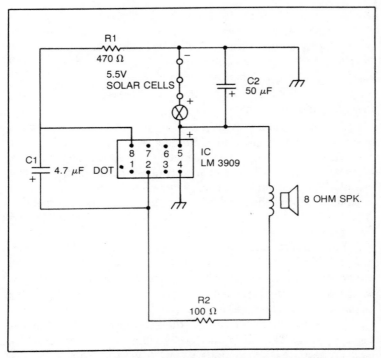

Fig. 23-5. Follow the circuit diagram to complete the wiring of all components. Notice the flashing LED is between the solar cells and pin 5 of the IC. Double check the wiring connections at least twice before firing up the black box.

Fig. 23-6. Mount the IC in the center of the perfboard. All other parts may be mounted around the IC. Push the leads of the capacitors and resistors through the holes and bend the terminal leads over.

181

positive terminal 5 of the IC. If the leads are reversed, the speaker will sound off with a constant tone and the LED will not light up. Now, reverse the LED leads. The LED should flash with a beeping tone heard in the speaker. Mount the perfboard in the plastic cabinet and hold in place with silicone rubber cement.

Preparing the Front Panel

The front panel may be metal or plastic. You may want to design your own speaker holes in the front cover. Here two large 1¼ inch holes were cut out with a metal circle cutter. The other two holes were drilled with a ½ inch bit (Fig. 23-7). The small speaker was placed on the metal plate and all four holes were marked. Draw around the sides of the speaker to give the speaker hole-guidelines. A ¼ inch hole at the bottom will let the red nose of the LED stick through.

First, mount the speaker with four 8/32 nuts and bolts. Next mount the LED. Mark on the metal plate the polarity of the LED for easy hookup. Place a dab of clear silicone cement over the back of the LED. Keep the LED leads upright. Let the cement dry before connecting to the perfboard circuits.

Solder the positive lead from the solar panel to the flat side of the LED (Fig. 23-8). Slip one inch spaghetti or larger insulation over the bare LED connections. Connect the other side of the LED

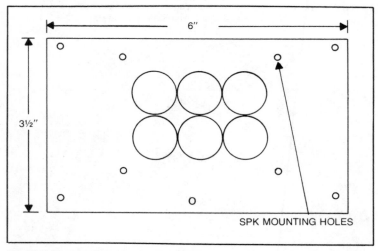

Fig. 23-7. The plastic or metal front holes may be cut with a circle cutter. You may design your own speaker holes. Place metal screen or speaker grill behind the small pm speaker.

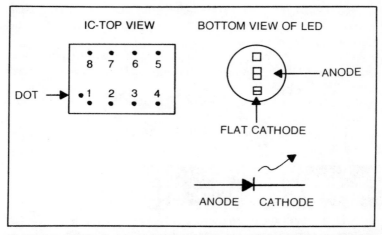

Fig. 23-8. Solder the positive lead from the solar panel to the flat side of the LED. If the speaker tone is constant, simply reverse the two leads to the LED. Correct terminal hookup is when the LED flashes and the beeping sound comes from the speaker.

to terminal 5 or the positive (+) terminal of C2. Try the unit out before replacing the front panel. Place a 100 watt bulb near the solar panel. Simply reverse the LED connections when a constant tone is heard without the flashing of the LED and beeping sound.

Troubleshooting the Black Box

When the speaker is dead and no flashing of the LED, check the voltage from the solar panel. Under a 100 watt bulb you should have over 5 volts applied to the IC. You may quickly check the voltage across pins 4 and 5 of the IC. No sound may be caused by a poor ground connection to pin 4. Check for a broken wire from R2 and pin 5 to the speaker terminals. Check all wiring connections.

Suspect improper wiring or a defective IC when the output solar panel voltage is very low. Remove the positive wire to the LED. Insert a vom in series with this lead and the LED. Set the vom to the 30 mA dc scale. This little unit pulls only 5 mA of current. When the current reading is 10 mA or higher, suspect improper wiring of the IC. Double check all wiring. A leaky IC may cause the high milliampere reading.

Here are five additional notes that may be helpful in building this blinking-beeping box:

☐ Bright light shining directly upon the flashing LED will make it flash faster. Keep direct light away from the LED.

☐ Less light on the solar cells will cause the LED to flash rapidly and lower the tone. Bright sunlight on the cells lowers the flashing and beeping sound.

☐ Less solar cells or lowering the voltage will develop lower volume and a faster flashing rate. It takes 5 volts to operate the LED normally.

☐ Adding more cells to the solar panel will make the LED flashing slow down.

☐ To lower the beeping sound, increase the resistance of R1. Decrease the resistance of R1 to increase the volume.

PROJECT 24
EXPERIMENTAL SOLAR CAR LIGHT REMINDER

How would you like to build a car light reminder that sounds off before you leave the garage? Many times the headlights are left on in the car after turning off the ignition key and going into the house. Obviously the next morning you know something is wrong when the car battery is dead.

Well, you can prevent this problem by constructing the solar car light reminder. If you don't want any more gadgets on your auto, this project is ideal. The head lights shine against the panel of solar cells and provide voltage to the solar alarm (Fig. 24-1). The alarm

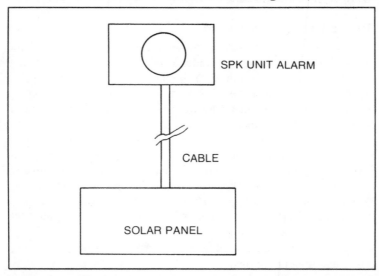

Fig. 24-1. The solar alarm is mounted above the solar panel. When the headlight beam strikes the solar panel voltage is applied to the alarm circuits.

Table 24-1. Parts List for Project 24.

```
IC — LM 386 — Radio Shack #276-1731
C1 — 22 mfd 16V electrolytic capacitor
C2 — 4.7 mfd 16V electrolytic capacitors
C3 — 22 mfd 16V electrolytic capacitor
R1 — 10K fixed 10 watt resistor
R2 — 10K fixed 10 watt resistor
R3 — 4.7K fixed 10 watt resistor
1   — IC 8 pin SOCKET — Radio Shack #276-1995
                        GC Electronics #F2-1000
                        GMI #12A41-540
1   — 8 ohm pin Spk.
1   — 3×3½" perfboard
12  — 100 mA Solar cells are greater
        Solar Amp. #S122, Poly-Paks, Inc. #5306
        H & R Inc. #TM21K666, John Meshna, Inc. #H-14A
        Edmund Scientific Co. #42,268
1   — Mini Experimenter Box 6¼ × 3¾ or Radio Shack #270-627
        GC Electronics #H4-722, ETCO #160VA
1   — 5 × 7 clear plastic photo frame
Misc. — Hookup wire, ac rubber cable
Construction time — 4 to 6 hrs.
Cost of Project —
```

lets out a loud wailing sound as long as the lights are left on. After turning off the motor, you can hear the alarm sound inside the auto. Thus, the headlights are never left on when the overhead lights are turned on with the automatic overhead door opener. This little project doesn't use up any energy, it develops it's own power. It has no batteries to run down or to replace either. See Table 24-1.

The solar panel is mounted down low where the full impact of one headlight shines on it. The overhead lights will not trigger the alarm since the car headlights must be within two feet of the solar panel. Since most garages are really too small, this is no problem. Mount the solar panel on the end of the work bench or wall where one of the headlights will shine against it. For best results, move the panel into the beam of the headlight. Usually, the car is parked in the very same spot since the car just has so much room to park with all the extra stuff stored around it. Greater voltage from the solar panel and more speaker volume may be obtained by placing metal reflectors around the solar panel.

How It Works

The sound-off alarm is constructed around a LM386 IC circuit (Fig. 24-2). There are no switches to operate the alarm. When the headlight beam hits the solar cells, voltage is applied to the IC component. The speaker sounds off with a loud honking or wailing sound until the voltage is no longer applied. Turning off the auto

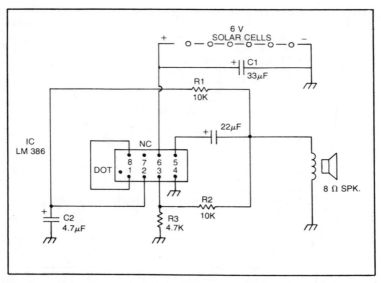

Fig. 24-2. The solar alarm is constructed around a LM386 IC circuit. A small 8 ohm speaker sounds off when voltage is applied to the circuit from the solar panel.

headlights, cuts the voltage from solar cells to the sound-off alarm. The tone of the speaker may be changed by decreasing C2 and R3.

Mount the speaker and all components except the solar cells inside a plastic 6¼ × 3¾ experimentor box. Cut a speaker opening into the metal or plastic front panel. If a round speaker is used, a circle cutter will do the job on metal or plastic. When a rectangular speaker opening is needed, cut several large circle holes in the front panel. Place the speaker on the front panel and drill out the required holes (Fig. 24-3). Place speaker grills over the speaker opening before mounting the speaker. The speaker is the only component mounted on the front panel. Drill a ¼ inch hole in the bottom of the plastic box for the solar panel cable.

All small parts are mounted on a 3½ × 3 inch perfboard. Cut the required size from a larger piece. The perfboard will mount in the bottom of the plastic box. Mount the 8 prong IC socket in the center of the perfboard. All other components are mounted around the IC as they are connected into the circuit. The capacitors and resistor leads are pushed through the small holes, then soldered to the various terminals (Fig. 24-4). Also, C1 is mounted on the perfboard. Only two wires from the speaker and two from the solar panel are soldered to the various connections on the perfboard.

186

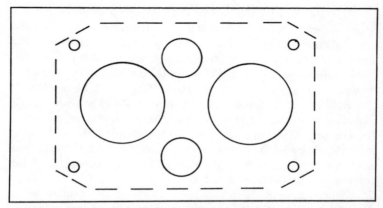

Fig. 24-3. Cut out the speaker holes after placing the speaker on the front panel. Place a speaker grill or screen wire over speaker before mounting. The size of the speaker may be any size, although the larger the speaker, the greater the sound.

The Solar Panel

The twelve solar cells will mount inside of a 5 × 7 clear plastic picture frame. Cut a piece of ⅜ inch foam material to just fit inside the frame. Mount the cells in the center of a piece of foam. Make three rows of four cells in a row. They will take up only 4 × 5 inches of the picture frame area. The clear plastic frame was used to protect the solar cells.

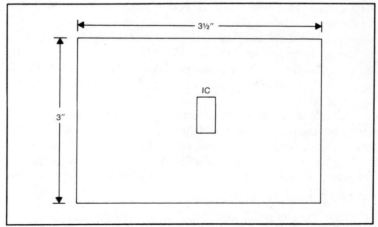

Fig. 24-4. All components are mounted on a 3 × 3½-inch perfboard. The components are wired and soldered together on the backside. Connect two wires from C3 and common ground to the small speaker.

Choose 100 to 150 mA solar cells for this project. One-fourth cells are ideal (Fig. 24-5). Although the alarm pulls only 25 milliamperes of current, the larger the cells the greater the volume. The current is greatest when more voltage is applied to the IC circuit. These cells are all connected in series to acquire 6 volts. Of course, if larger cells are used, a larger plastic frame may be needed.

Solder a two inch piece of hookup wire (Radio Shack wrapping tool wire #30 gauge Kynor is ideal), to the bottom of each cell. Draw a line where the cells are to be mounted. Line them up together. Place a dab of clear silicone rubber cement to the bottom of each cell and stick it in place. Bend the wire upward until the cement sets up in about four hours.

Now, scrape the insulation back or burn off the insulation with the iron and connect the wire to the top center bar of each solar cell. Solder a 10 inch length wire to the top of cell number 1. Likewise, solder or connect a 6 inch wire to the bottom side of cell 12. These two wires will connect to the ac cable connecting the solar panel to the alarm box.

Place a dab of rubber cement in each corner of the plastic frame and to each corner next to the solar cells. This will prevent damage to the cells as they are back away from the front plastic frame. Mount the cells into the plastic frame. Afterwards, seal the backside of the foam to the plastic framework. This will prevent dust and dirt from getting inside to the cells.

Use a pocketknife and cut out a small area of the foam in the back for the cable connection. A piece of flat rubber ac cord is ideal to connect the two units together. The length of the ac cord will depend on what distance the alarm is mounted above the solar cells. Solder up the two leads to the ac cable. The solar panel polarity may be marked when a voltage reading is taken with the vom. Mark the negative terminal with a black felt pen and the other brown wire as the positive terminal.

Check the solar panel out before sealing up. Place a 100 watt bulb over the panel. You should measure 5.5 to 6 volts at the end of the ac cable. Double check each cell and wiring connection for any voltage lower than 5 volts. Visually inspect each cell for a crack or possible breakage. Replace the broken cell and take another voltage test with the vom.

Testing

Cut the solar panel cable the required length between panel and alarm. The two units may be connected together and tested

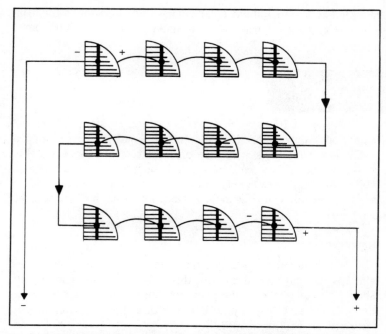

Fig. 24-5. Here are 125 mA one-fourth solar cells. Mount the cells four in a row. Connect all cells in series.

before mounting in the garage. Double check the polarity of the panel cable. Make sure the positive voltage wire goes to the (+) terminal of C1. The negative wire goes to common ground. Now, place a 100 watt bulb over the solar panel. The alarm will sound off when the bulb is within ten or twelve inches of the solar panel.

In case the alarm is dead, check the cable polarity. Take a voltage check across C1 and at the solar panel connections. The solar panel should not be sealed with rubber silicone cement until the alarm is functioning. Check for 5 or 6 volts at pin 6 of the IC. Check for a good ground at terminal 4. Place the ohmmeter leads across the speaker and if okay, you should hear a clicking noise. Make sure the light bulb is off when making resistance checks. Usually, double-checking wiring and cable connections will uncover the defect.

Now, install the solar alarm in the garage. The alarm box should be mounted up high where it is easily heard. Mount the solar panel in front of the headlights. Several position tests may be necessary for correct beam location. The headlights must be within two feet of the solar panel. More voltage and greater volume may be obtained with more solar cells added in series. To protect the solar panel, mount a 2 × 4 framework around it so the car will not bump against it.

This solar IC radio circuit is actually a crystal set front-end amplified with a simple IC audio output. The little AM radio provides adequate speaker volume on local broadcast stations (Fig. 25-1). A total of only ten separate components are found with a six volt solar cell power supply. Fifteen solar cells wired in series or six separate one volt cells may be used in the solar panel. See Table 25-1.

How It Works

With a good outside antenna, local and distance broadcast stations may be picked up with this solar IC radio. The antenna should be connected to the primary winding (1) for distance stations. Local stations may be heard with a throw-around antenna wire attached to terminal 2.

The antenna picks up the rf broadcast signal and is tuned in with L1 and C1. The separated stations are detected with a crystal diode D1. R1 controls the weak rectified audio signal on pin 2 of IC LM386. IC1 amplifies the weak audio signal and is coupled to the 8 ohm speaker with C3 (220 μF). The solar cells provide a dc voltage to the output power IC for audio amplification.

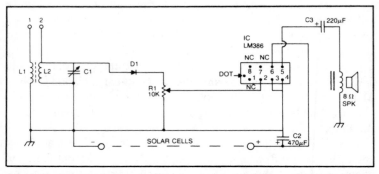

Fig. 25-1. The little AM radio provides adequate speaker volume with local broadcast stations. Only ten separate components are needed besides the solar panel.

190

Table 25-1. Parts List for Project 25.

L1 & L2 — Ferrite antenna coil with 32 turns over one end
    (See text).
C1 — 365 vfd variable capacitor
    GC Electronic #A1-233
    ETCO Electronics Corp — #185 VA or 042 CC
C2 — 470 µf 16V electrolytic capacitor
C3 — 220 µF 16V electrolytic capacitor
R1 — 10K variable volume control
IC — LM 386 — Radio Shack #276-1731
Socket — 8 pin IC socket — Radio Shack #276-1995 or equiv.
SPK — 8 ohm 2 TV 4 inch pin Spk
15 — Solar cells — 100 mA, Solar amp #S122,
    Poly-Paks, Inc. #5306
    H & R, Inc. #TM 21K666
    John Meshna, Inc. #H-14A
Misc. — 1 inch foam, wood, masonite, screws, etc.
Cost — $40.00
Time — 6 to 7 hrs.

Preparing the Chassis

Of course, any type of cabinet or mini-box may be used to house the small solar IC radio, except here an "L" type chassis was constructed (Fig. 25-2). The front panel consists of a piece of ¼ inch masonite. Pick up a piece from the scrap box or take a section from the back or sides of an old TV set. The sides of an old TV cabinet is ideal for most masonite surfaces have a wood type finish. A piece of ¾ inch plywood or solid pine wood works nicely for the bottom chassis.

Since the fifteen solar cells take up six inches of the chassis surface, the front masonite panel should be cut 5 × 8 inches. The rear or bottom chassis dimensions are shown in Fig. 25-3. Sand down all four edges of the piece of masonite and wood chassis. After

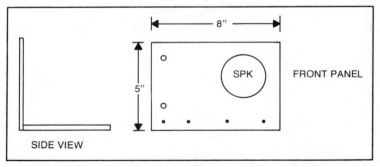

Fig. 25-2. A piece of masonite and wood form the chassis. All panels are drilled and sanded before a finish of lacquer or varnish is applied.

all holes are drilled, both panels are sprayed with clear lacquer or varnish.

Center and mark the ⅜ inch holes in the front panel. The volume control and tuning capacitor are located 2 inches from the left end of the panel. Drill four ⅛ inch holes at the bottom edge to secure the front panel to the wood chassis. The speaker hole may be cut out with a jig or coping saw or several ¼ inch holes drilled in the speaker area. If the area is cut out, place a piece of screen to protect the speaker. No holes are drilled in the wood chassis.

The Antenna Coil

Choose a ferrite antenna rod if you are going to roll your own coil. Start about ½ inch from one end and wind a total of 72 turns with number 26 enameled wire (Fig. 25-4). Actually, number 22 to 28 enameled wire may be used here. You can salvage this wire from other coils, old transformer or old speaker coils, or relays. The actual wire size may be picked up at most radio-TV, repair firms.

Secure each end of L2 with cellophane or scotch tape. Now, over one end of L2 wind another coil of 32 turns. Wrap cellophane tape over coil L1 to hold the wire. Leave about 8 inches of wire extending from each coil. You may have to add hookup wire at the extreme end of L2 to connect to the ungrounded terminal of C1. The ferrite coil may be set aside until mounting.

Connecting the Parts

Mount the variable capacitor and volume control on the front panel. The tuning capacitor is mounted above the volume control. Bolt or cement the speaker in place. Tip the chassis over on its face

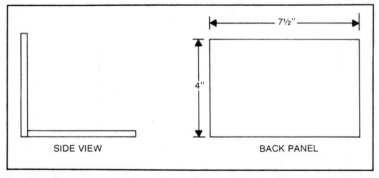

Fig. 25-3. Cut the back chassis from a piece of plywood or ¾ inch pine. The two panels will be held together with four small round head wood screws.

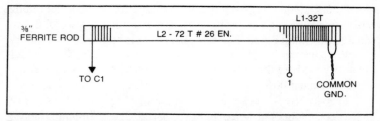

Fig. 25-4. Select a 6 inch ferrite rod ⅜ inch in diameter and wind 72 turns of number 26 enameled wire. Over the other end, wind 32 turns of the same type of wire for L1.

and mount the antenna coil L1 and L2. This coil may be bonded to the masonite panel with rubber silicone cement at each end of the coil (Fig. 25-5). Now, let the cement set up for at least four hours.

After all major components are mounted, the smaller components may be connected. First, solder the long end of L1 to the ungrounded terminal of C1. Scrape and tin the enameled wire for good soldered connections. Twist the bottom end of both coils together and solder to the grounded section of the variable capacitor. Form a short loop in the open end of L1. Likewise, form a loop in a piece of hookup wire and solder to terminal 2. These two coil connections are held with a dab of rubber silicone cement to the front panel. A small alligator clip may be used to connect the antenna to either antenna terminal.

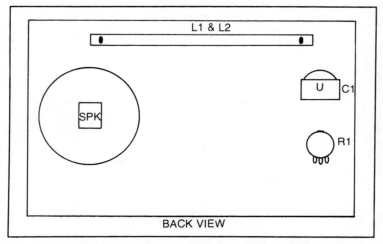

Fig. 25-5. The coil is located at the top of the masonite panel. Bond the coil to the front panel with rubber silicone cement at each end.

Solder diode (D1) from the ungrounded terminal of the tuning capacitor to the high side of the volume control. Be careful not to damage the small diode. For safety, use a pair of needle nose pliers as a heat sink while soldering the diode into position. Now, solder the bottom side of the volume control to common ground. Later the center terminal of R1 will connect to terminal 2 of the IC socket.

Prepare the IC socket before mounting (Fig. 25-6). Solder an 8 inch piece of hookup wire to terminal 2. Tie terminal 3 and 4 together with a common or ground piece of hookup wire. This wire may be left bare so other components can be soldered to it. Connect C3 to terminal 6 of the IC socket. Now, solder a ten inch piece of hookup wire to pin 6.

Before connecting the IC socket to the other components, double check each connection. Now, mount the IC socket close to the front panel and right under the volume control. Place a dab of silicone cement on the board chassis and stick the socket in place. Bend C3 terminals and all connecting hookup wire so the socket will lie flat. Apply rubber cement to C3. Let the cement set up overnight.

Finish connecting C3 (200 μF) to the 8 ohm speaker. Use hookup wire to extend the lead of C3 and ground connection. All electrolytic capacitors may be bonded to the wood chassis with rubber cement. Solder the hookup wire from terminal 6 to the positive side of C2 (470 μF). Ground the other end of C2 to the bare

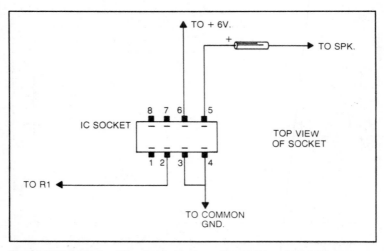

Fig. 25-6. Prepare the IC socket before mounting. Solder all leads and components to the socket.

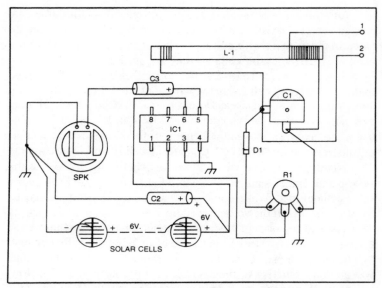

Fig. 25-7. Follow and check off each component on the wiring drawing.

chassis ground wire. Before going any farther, check the connections at least twice. Follow and check off each component in Fig. 25-7.

Building the Solar Panel

Select fifteen solar cells with 50 to 150 mA of current capacity. Or you may use six one volt solar cells. Also a ready-built 6 volt solar pak may be used here. These cells will mount on a one inch piece of foam material (Fig. 25-8). Usually, the 1 volt cells will take less space than two 150 mA cells. First, place the total number of cells on the piece of foam and cut the dimensions accordingly. The foam material may be cut with a pocketknife or regular saw.

All solar cells must be connected in series to acquire 6 volts. If the bare cells do not have any leads, solder a piece of wire to the bottom side of each cell. The last cell should have a 10 inch piece of wire soldered to the bottom side. Connect a 10 inch piece of wire to the top side of cell number 1. Often, one volt cells will have a red and black wire attached. Just connect the red wire from one cell to the black wire of the next cell until all cells are wired together. Of course, each wire must be cut to the right length before soldering. The black wire connects on the top of the cell with the red wire to the bottom (+).

After all wires are soldered to the bottom side of the bare cells, place a dab of clear silicone rubber cement and place in line on the piece of foam. Start with cell number 1 at the left and back of the foam area. Arrange the cells so they are in line (Fig. 25-9).

Now, cut each wire lead so the bottom wire from cell 1 will go to the center bar of cell 2. Tin the wire, if needed. Lightly solder the wire to the center bar area. Do not press down too hard upon the cell. Try to make a neat soldering connection. If too much heat is used, the heat may loosen the cement or the silicone material may pull up from the top of the cell. Connect all cells in the same manner.

When all cells are wired in series, check the voltage of the solar panel. Under sunlight or a photo light, you should have over 6 volts without load. While under a 100 watt bulb, the voltage may be from 4 to 5 volts. If the voltage reading is below these figures, check for a cracked cell or poor soldered connection. Measure the voltage across two cells (.75 to 1 V) until you have found the defective one.

Connect the solar panel to the IC radio. The positive terminal will go to the positive terminal of C2 and terminal 6 of the IC. Now, insert the IC into the small 8 pin socket. Check the dot on top of the IC for terminal number 1. This dot will plug in over terminal 1 of the IC socket. Double check as you may ruin the IC if plugged in backwards. Solder the negative wire of the power panel to the bare ground chassis wire.

The IC radio is now ready to be checked out. Place a 100 watt bulb over the solar cells or lay them in the sun. Rotate the volume control wide open. Connect the antenna wire to terminal 2. Now

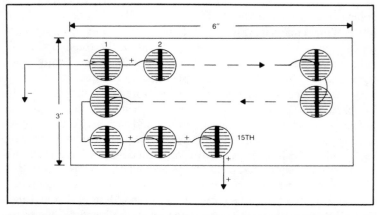

Fig. 25-8. Mount the solar cells on a piece of 1 inch foam material. All solar cells are connected in series.

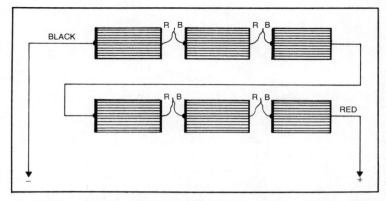

Fig. 25-9. Here six 1 volt cells are wired in series. These cells take up a little less room than a total of 15 100 mA cells.

rotate the tuning capacitor until a station is tuned in. Distant stations with better tuning separation are made with a long antenna connected to the terminal of the antenna coil. A good ground connected to the common ground terminal may help the volume on distant stations. Check for your local stations. The volume control should be turned down when one is located.

If no station can be received, check the audio amplifier for a hum or click at the top of the volume control. Place a screwdriver blade on the volume control. You should hear a hum or click. If not, remove the top wire of the volume control and touch the connection on the control. A hum should be heard. You may hear a motorboating noise under a 100 watt bulb. Check the voltage at pin 6 of the IC. Now go to the diode and rf section. Check the diode for breakage. A shorted diode will pass current in both directions. Check all wiring connections to the antenna coils (L1 & L2). It's possible one of these small wires may have broken off. Check the coil wires for good bonding. Improper tinning and scraping of the enameled wire may result in a poor coil connection.

Although the speaker will not blast your ears, the volume is adequate for a small solar IC radio. The amount of stations selected and received will depend upon the long antenna and the number of stations in your area. One thing for sure, you never have to worry about turning the radio off. Just set the rear of the receiver in the sun and enjoy.

Chapter 4

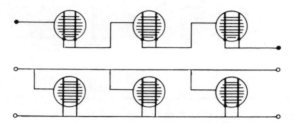

Five Solar Projects for Under $50

**PROJECT 26
SOLAR POWERED TOY AUTO**

You can now purchase toy autos, trucks, and vans that are powered by solar cells (Fig. 26-1). Some of these kits are ready to rur. While in others, the motor, pulley, and gear assemblies must be mounted separately. Other motor kits may be used to mount in beanies, airplanes, or small boats. Let your imagination go since you can solarize most toys.

Usually, these small solar toys may be powered by one large solar cell. Of course, it will depend upon the weight of the toy. Do not expect to power a large or heavy toy with these solar kits. Most solar kits have a 1.5 volt dc motor for solar power. It's best to choose a small motor operating up to 1.5 volts with 150 to 200 mA of current. See Table 26-1.

Choosing the Solar Cells

Most solar toy kits include one round 2″ solar cell. More than one cell may be added to produce greater speed and power. Choose solar cells (for small motor operation) of .500 amps or greater. Usually, these cells are 2 to 3 inches in diameter. You may connect two .250 amp solar cells in parallel to acquire the correct current.

If only one cell comes with the solar toy and you desire more speed and power, connect the cells in series and parallel to get the

Fig. 26-1. Here is a fire truck equipped with solar power. Some of these kits are assembled and ready to run.

required operating voltage and current. If you do not have another 2″ cell, but a couple of 250 mA or greater half round solar cells, just connect the two in parallel and then in series with the round cell (Fig. 26-2). To operate a separate motor connect four 250 mA cells in a series parallel arrangement (Fig. 26-3).

Table 26-1. Parts List for Project 26

Solar Truck and Van Kits —	Silicon Sensors, Inc.
	Highway 18 East
	Dodgeville, WI 52533
	Solar Electric Eng. Co.
	435 West Cypress St.
	Glendale, CA 91214
Solar Motor Kits—	#42741
	Edmund Scientific Co.
(1)	101 E. Glouschester Pike
	Barrington, NJ 08007
(2)	Experimentor Kit —
	Silicon Sensors, Inc.
	Highway 18 East
	Dodgeville, WI 52533
(3)	#103
	Solar Power Cell and Motor Kit
	Solar Electric Eng. Co.
	438 West Cypress St.
	Glendale, CA 91204

Construction Time — 2 hrs.
Cost of Project — Kits $8.00 to $30.00

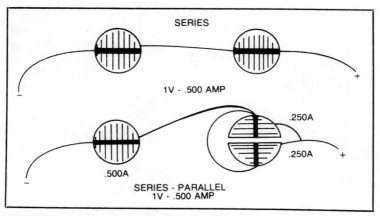

Fig. 26-2. Add one or two more solar cells for greater power and speed. Connect two 250 mA cells in parallel and then in series with the larger cell.

One must remember that these small toys will run only upon a smooth surface. With adequate cells (2 or more) the small solar powered toys will scoot rapidly on a cement sidewalk. Use at least two cells to run up a small incl ne of a cement sidewalk. Of course, the speed going uphill is quite slow, but they really roll going downhill.

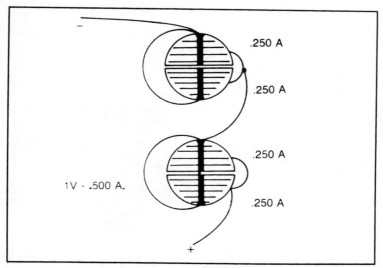

Fig. 26-3. If only 250 mA cells are handy, connect them in a series parallel arrangement to acquire 500 mA at 1 volt. Usually, the smaller cells may be mounted anywhere.

The solar cells mounted on top of the toy truck or car may be tested with a 100 watt bulb. This bulb has to be quite close to the cells to operate the small motor. Use the 100 watt bulb for testing only. *Do not use line cords on bulbs with boat toys around water*. Lift or elevate the powered wheels for these tests.

When the small wheels will just barely move or not rotate at all, suspect poor wiring connections or the solar cells are too small in current rating. First, check all wiring connections. If the connections are good, make a current check. Simply take one lead off the solar cell and insert a milliampmeter lead between the wire you removed and the solar cell (Fig. 26-4).

Now, place a 100 watt bulb on top of the solar cell. Set the meter on the 300 or 500 mA range. If the current is greater than 250 mA, either the motor is shorted or pulls too much current. Connect a flashlight battery to the motor to check it. If the motor rotates, the motor is normal, but it just will not operate off the solar cells. Choose a 1.5 volt motor with an operating current around 150 mA.

Adding More Cells

Here, two more 2¼″ solar cells were added to the fire truck for more power. They are easily mounted on the top of the ladder racks (Fig. 26-5). Not only will it increase the voltage to 1.5 volts for

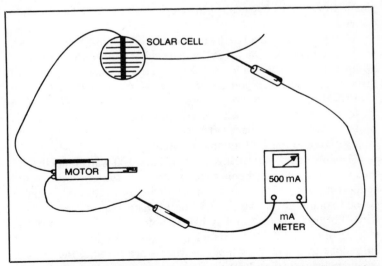

Fig. 26-4. To determine what current the motor is pulling, place a milliampmeter in series with one motor lead and the solar cell. Set the meter on the 300 or 500 mA range.

Fig. 26-5. With this fire truck, two more 2 ¼ inch round solar cells were added for power and speed. Now the fire truck will even run up a hill.

speed, but will let that truck run up a slight incline of the sidewalk while in the sun.

Remove the toy's cap housing of the solar cell. This cell will pop right out of the plastic container. Unsolder the black wire from the motor. Make a hole with the soldering iron tip through the plastic towards the back cab of the truck. Now run the black wire to the red wire of the next cell.

Connect a long lead to the motor terminal from which the black wire was removed. This wire should be about 8 inches long and may be cut to length as you solder up the last cell. Be sure to connect all three cells in series. If you are mixing cells, make sure the top side of the first cell connects to the bottom side of the next cell. Some of these cells have the red wire connected to the top side of the cell while others have the black wire. Remember the top side of the cell is negative and the bottom side positive.

Double check the wiring. Now, place a 100 watt bulb close to the cells. Prop up the wheels so the motor will run. The light should be almost on top of the cells to make the motor rotate. If the motor doesn't rotate, check out the wiring. Perhaps one of the soldered wires got broken off. You may have a defective cell. Measure the voltage of each cell. You should have close to 5 Vdc with the light bulb close to the cells. When the wheels are turning, the voltage should be from 1.25 to 1.5 Vdc. Reverse the two motor wires if the wheels are turning backwards.

Now, tape up all bare connections. Masking tape may be used here or spaghetti insulation over the soldered connection. Since we are working with low voltage, rubber or plastic tape is not required. Pull the wires down under the two added solar cells. Keep the wires in the center of the truck bed.

Place a dab of clear rubber silicone cement on each side of the cell. Dab up both areas where the cells will rest. Now, mount the two cells. Hold them in place with a heavy object. A paperweight or stapler will do nicely. Now, wipe off the excess cement. After the cement sets up overnight, additional cement may be added underneath the cells so they will not become dislodged in a truck accident.

Separate Motor Kits

There are several motor replacement kits that you can use to motorize any small plastic toy. These kits consist of two sets of plastic wheels and axles. You will find separate gears for mounting the motor to the axle or small gears that may be attached to the original set of wheels. Some kits furnish bushings to build up the gear assemblies, if need be.

Simply mount the motor and gear assemblies. Hold them in position with airplane glue or rubber silicone cement. The motor may be cemented to a block of wood or styrofoam to raise the motor into position. Now, glue the block of wood or styrofoam to the plastic toy.

Larger toys may be powered by one or two motors, one in front and the other in the back. Each set of wheels will be power driven. Connect the required solar cells in series and parallel to provide the correct voltage and current. Usually, in larger trucks and autos, there is more room for mounting extra solar cells.

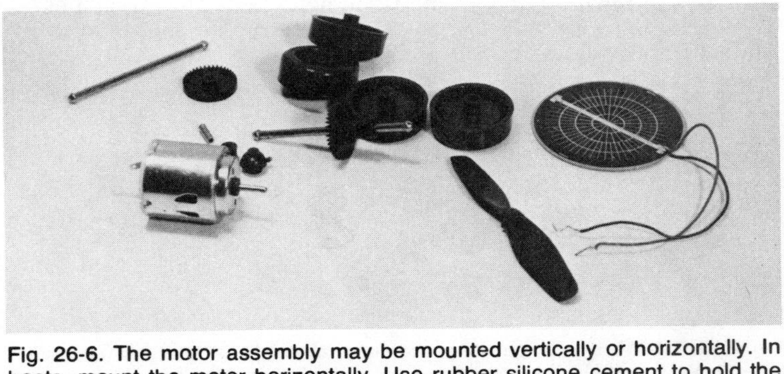

Fig. 26-6. The motor assembly may be mounted vertically or horizontally. In boats, mount the motor horizontally. Use rubber silicone cement to hold the motor in place.

Why not build a couple of small solar powered boats and let them race in the water. Simply mount the motors horizontally with the shaft out the back (Fig. 26-6). Mount the airplane propeller and watch them go. After completing any toy with rotating wheels, a drop of light oil on the plastic bearing keeps the vehicle moving.

PROJECT 27
SOLAR KID GAMES

Your new computerized electronic football game lets you and your opponent match your skills against an internal computer. Decide whether to pass or run and the computer plays defense. This fast-paced game has touchdowns, field goals, punts, interceptions and all the excitement of a real football game. Most games of this type are operated from a 6 or 9 volt battery. These batteries may last for a long time, but with constant use and if the switch is left on overnight, you have to purchase another battery (Fig. 27-1). Of course, they always seem to quit during the best times.

Simply build a 6 and 9 volt regulated solar-pak to prevent this costly situation. You have no batteries to wear down or switches to turn off. Those solar kid games can be played during the day on solar power and at night with batteries. Of course, some of these hand-held games may be played with the solar panel under a 100 watt reading lamp. See Table 27-1.

Games with excessive current drain cannot be operated with the solar-pak. First, check the operating current and voltage of the electronic game. Forget building the solar-pak, if the operating current is over 100 mA. Most electronic kid games pull from 10 to 50 mA of current. Usually, the voltage is either 6 or 9 volts. This solar-pak is ideal for most electronic hand-held playing games.

Fig. 27-1. There are many games or projects that may be powered with this solar project. You don't have to worry about batteries going dead or leaking. In fact, you don't even have to turn the game off!

Table 27-1. Parts List for Project 27.

20 100 to 250 mA solar cells — Poly Pak #5312
     Edmund Scientific Co. #42,268 — John Meshna, Inc. #H-14A
     H & R Inc. #TM21K932 — Solar amp #S124
2 jacks — Shielded phono jack — Radio Shack #274-346 or equivalent
2 plugs — Shielded phono plug — Radio Shack #274-334
D1 — 6.2 V zener diode — Radio Shack #276-561 or equivalent
D2 — 9.1 V zener diode — Radio Shack #276-562 or equivalent
1 — 5×7 clear plastic picture frame
Misc. ⅜ inch foam material, solder, hookup wire
Construction Cost — Under $50.00
Time — 4 to 6 hours

Checking the Current

To determine the voltage the small games use, check the owner's operating manual or look inside the battery compartment. The voltage is listed either on a piece of paper or stamped right into the plastic. Count the batteries when the voltage chart cannot be located. Small "AA", "A", or "C" batteries are 1.5 volt cells. Multiply the number of cells times 1.5 and you come up with the operating voltage. For instance, if there are four "AA" cells, you know the operating voltage is 6 volts. Often when a single large cell is used, the operating voltage is 9 volts.

Now, determine the operating current. When only a 9 volt battery is used, leave one clip out and turn the battery half way around (Fig. 27-2). Insert a milliampere meter between the two

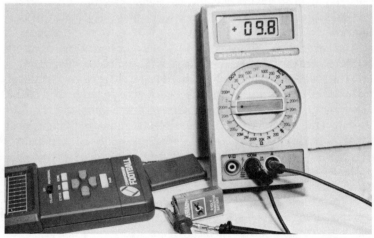

Fig. 27-2. Measure the operating current of the game or project with a vom. Insert the leads of the vom between battery connections and terminal. This hand-held football game pulls about 10 mA of current.

contacts to read the current. Turn the game on and notice what maximum current is drawn. This will determine whether the game can be played with solar power. If four small batteries are used, insert a piece of cardboard between the positive cell terminal and contact. Touch the current meter leads to the positive battery terminal and metal contact. Now, read the operating current. This 6 and 9 volt solar-pak will operate most electronic kid games pulling less than 50 mA of current.

Connecting the Solar-Pak

The electronic hand-held game may be connected to the solar-pak with a headphone jack and plug. The game can still be held with one hand and operated from solar power. Simply unplug the jack when not in use or if you desire to use battery power. A small female earphone jack is installed at the bottom of the electronic game. Installing and soldering the jack may be easily done through the battery compartment.

Remove the plastic lid and battery. Check each side of the plastic case. Usually, there is plenty room to install a small earphone jack. In some cases you may have to remove a small piece of plastic from the battery compartment to install the jack. Do this with the tip of the soldering iron. The side hole for the headphone or phono jack may be pushed in with the tip of the soldering iron. Be careful not to make the hole too big. Cut off the excess edges with a pocketknife. The hole may be drilled, being careful not to damage other components. Connect the two wires to the jack before mounting.

In case a 9 volt battery is used, simply connect a 9 volt battery terminal harness to the headphone jack. The harness will plug right into the present battery cable plug. Always, observe correct polarity. The red wire is positive and the black wire negative. If in doubt, check out the continuity of each wire with the ohmmeter. Look at the top of the 9 volt battery for the correct terminal markings. Most 9 volt batteries are marked at the top, alongside the terminals.

When single batteries are used, locate the wires going to the positive and negative terminals of the battery contacts. Generally these wires are color-coded, red and black. Run the positive or red lead to the end terminal of the headphone jack (Fig. 27-3). The common or ground terminal of the jack and plug is negative. Double check for correct battery polarity. Leave the small batteries out when operating from solar power.

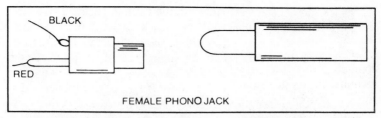

BLACK

RED

FEMALE PHONO JACK

Fig. 27-3. Connect the red lead to the center terminal of the headphone or phono jack. The common or ground terminal goes to the outside or shielded area of the jack.

The Solar Panel

Select twenty 100 to 500 mA solar cells. All cells are wired in series. A 6.2 and 9.1 volt zener diode is used for 6 and 9 volt regulation (Fig. 27-4). Mount the solar cells and diodes in a 5 × 7 plastic picture frame. Two-conductor speaker wires connects the solar panel to the solar kid game.

Cut a piece of foam material so it will fit inside the plastic frame. Mount the solar cells toward the center of the foam material. Leave a 1 inch clearance on each side of the frame. For each corner of the plastic frame, four suction cups will hold the plastic picture frame to the outside window pane. These suction cups should be of the smaller size so not to shield out any light to the solar cells. Pick them up at a hardware or retail lumber store.

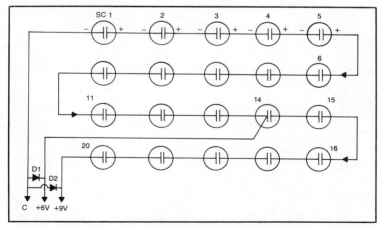

Fig. 27-4. All twenty solar cells are wired in series. A 6.2 and 9.1 V zener diode is used for 6 and 9 volt regulation.

Since the cells are cemented to the foam material, a two inch piece of wire should be soldered to the back of each cell. Place a dab of clear rubber cement on the back of the cell and push it into the foam material. Be careful not to break or crack the cells. Keep all cells in line. After the cement has set up, connect the wire to the bar area of the next cell. You are now connecting a positive lead to the negative bar on the top of each cell. Solder an 8 inch connecting wire to cell 1 and cell 20.

Push the two connecting wires through the foam material. Use an ice-pick or tooth-pick to make the small holes. For correct color-coding, solder a black wire to the top side of cell number 1 and a red wire to the back side of cell number 20. These two wires are pulled down to the cable connections. Count from cell 1 to cell 14 and connect a tap wire to the top of this cell. This will be the 6 volt tap-off connection. Feed the wire through the foam area.

Solder the 6.2 V zener diode to the connecting wire of cell 14 and common ground. The positive or white ring marking of the diode must be connected to the positive terminal. Solder the negative or anode terminal to the common black wire. Likewise, solder the 9.1 V zener diode to the red connecting wire of cell number 20 (Fig. 27-5). Double check the polarity of each diode. These diodes may be wired across the corresponding voltage jack.

Before mounting the solar panel inside the plastic frame, drill all holes and mount all components on the framework. Drill four ⅛

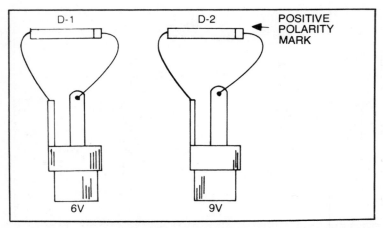

Fig. 27-5. Solder the two zener diodes across the respective shielded jacks. Check for the correct polarity. The negative terminal (anode) goes to the shield or outside jack terminal. The positive terminal (cathode) solders to the center pin of the jack.

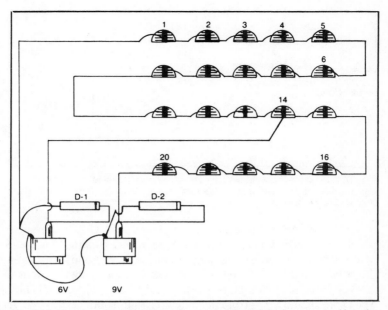

Fig. 27-6. Here is a diagram of how the solar cells and diodes are wired into the circuit. Observe the correct polarity of each zener diode.

inch holes in each corner of the plastic frame to mount the rubber suction cups. Drill two ¼ inch holes in the bottom center area of the plastic frame. Mount two shielded phono jacks in these holes. Select the flush jack which accepts the RCA type phono plug. Use long metal phono plugs, so they can easily be removed in each end of the cable. In fact, two wire speaker cable is ideal to connect the solar panel to the hand-held solar game.

Mount the suction cups and shielded jacks before the solar panel is inserted. If the holes of the jacks are drilled in the center edge of the plastic frame, only a small amount of foam must be removed around the voltage jacks. Check the solar panel before the panel is installed in the plastic frame. You should have 6 and 9 volts, when in the sun, and 5 and 8 volts under a 100 watt bulb. If lower voltages are noted under the sun, check the cell connections. A poor connection or cracked cell may lower the output voltage.

Mount the foam panel inside the plastic frame. Place a dab of clear silicone rubber cement in each corner. This will prevent damage to the solar cells. Connect the common wire (black) to the shielded or outside area of the metal jack. Solder the 6 and 9 volt leads (red) to the center jack terminals (Fig. 27-6). Solder each

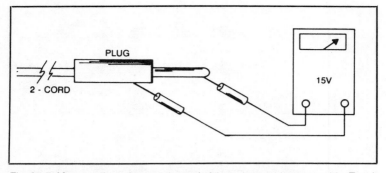

Fig. 27-7. Measure the voltage at the end of the solar panel to wire cable. Touch the positive lead of the vom to the center plug terminal. If the meter goes backwards, reverse the two leads.

zener diode across the respective voltage jack. Check the solar panel for correct voltage with the vom before seating the foam material in place. Cut another piece of foam material and insert over the solar panel. Seal all edges with clear silicone rubber cement.

Connect two male plugs to each end of the connecting cable. The cable should reach from the plug area to the outside window. A cable that is too long may lower the solar panel voltage. Solder the negative terminal to the outside shield and the positive connection to the center plug terminal. Insert the male plug into the solar panel and check the voltage at the other end of the plug for correct polarity and voltage. The positive terminal of the vom should be touched to the center plug terminal (Fig. 27-7). If the meter hand goes backwards, the two wires should be reversed. When the correct voltage and polarity are measured at the male plug, insert it into the solar kid game and enjoy. You may also find the solar panel will supply voltage to many other games, electronic kits, and experimental projects.

PROJECT 28
SOLAR BIKE RADIO

How would you like to own a solar bike radio and never have to worry about charging batteries or having the batteries leak? With this solar bike radio you may listen to music as you take that leisurely ride or do those youthful errands. The small bike radio will operate even on cloudy days. This bike radio operates from *solar power*. A commercial solar panel may be used if you do not want to construct your own (Fig. 28-1).

Fig. 28-1. Here is the small bike radio with a commercial panel. Cement the solar panel to the top of the bike radio.

Choosing the Radio

There are a number of bike radios on the market and some even come with a new bicycle. Choose a bike radio with a current operating capacity of less than 25 mA. Another added feature is one that operates with low battery voltage. We choose the Radio Shack AM bike radio since it operates with three "C" batteries (4.5 Vdc) and pulls only 20 mA at full volume. See Table 28-1.

Another feature to consider when selecting the bike radio is a flat plastic surface. Slanted areas may cut down the reception when the cells are cut off from the sun. So choose one with a fairly flat top area. One with a metal surface will do since a plastic box may be constructed to house them, but with this radio, the solar cells are cemented directly on the plastic top area of the radio.

Table 28-1. Parts List for Project 28.

1 — Bike radio — Radio Shack 12-197 ($19.95)
11 — Solar cells (25 mA or greater) Solar Amp #S122, Edmund
    Scientific Co. #30,735 or (crescent cells P-42,749),
    H & R Inc. #TM20K187, John Meshna, Inc. #H-11,
    BC Electronics #J-40800, Poly Pak, Inc. #5538 or 5300.
Misc. Piece of foam material, plastic box lid, solder, etc.
Commercial Solar Panel — Solar AMP S-116 #(3,6,9 Vdc),
    Edmund Scientific Co. #42,745 (60). (A commercial solar
    panel may be used if you desire not to build one).

Project cost — Under $40.00 (including bike radio)
Construction Time — 6 to 7 hrs.

Voltage and Current Checks

Usually, the service literature or sales catalog will indicate what voltage the small radio uses. If not, simply count the small batteries and multiply by 1½. The batteries will indicate the operating voltage. This small radio operates with 3 "C" cells, indicating a total voltage of 4.5 Vdc (Fig. 28-2). When only one battery is found, check the battery for voltage identification. Only one battery may indicate a 9 volt type.

After determining the operating voltage, check the current rating. Most small battery radios have an operating current capacity under 25 mA. Make sure and check it out. With this radio, a piece of cardboard was inserted between the positive terminal of the battery and the contact. Pry the battery back and slip a piece of cardboard between the two contacts (Fig. 28-3).

Set the vom current range at 30 mA. Touch both contacts with the meter probes. The meter hand may go backwards, but the radio should operate with the switch on. Reverse the meter leads for correct current indication. The meter hand will indicate in the right direction when the positive terminal of the meter probe is connected to the positive terminal of the battery.

Now, turn the radio on and tune in a loud local station. The radio will pull maximum current with a station tuned in. If the meter hits the peg, the radio pulls over 30 mA and the next current scale should be used. Actual current test with the bike radio was 20 mA at 4.5 Vdc operating voltage. After determining the current and voltage rating of the radio, you may now design a solar panel to operate the radio.

Fig. 28-2. Check the operating voltages by counting the batteries. Only one battery may indicate 9 volts. Check the battery body for the correct voltage.

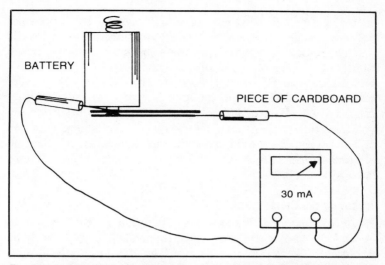

Fig. 28-3. Insert a piece of cardboard between the positive terminal and contact. Clip the current meter leads to these two contacts and read the operating current with the radio turned on.

Choosing the Cells

Practically any 25 mA or higher solar cells may be used, since the small radio pulled only 20 mA of current. These small solar cells come in all shapes and sizes, half round, quarter, crescent, or just small broken pieces. Here we choose .125 amp ¼ round cells.

Eleven solar cells were chosen to operate the small radio (Fig. 28-4). Although, the total voltage would be around 5.5 Vdc, the radio

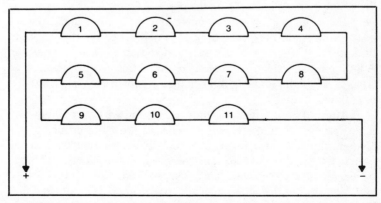

Fig. 28-4. Connect the eleven solar cells in series for a 5.5 Vdc output. This will insure normal operation even on a cloudy day.

should operate efficiently even on cloudy days. So, two more solar cells were added in series. Not only is the selection of solar cells important in current and voltage ratings, but cells should be chosen to fit upon the top of the small radio surface.

A total of thirteen cells will operate a 6 volt radio, while twenty cells are needed for a 9 volt operation. Notice if the correct amount of cells will fit on top of the radio, after determining the current and voltage rating of the radio. The eleven solar cells used with this radio left a ½ inch exposed area at the top. Remember, crescent and ¼ inch solar cells may be mounted close together and not take up a lot of space.

Mounting the Cells

Before mounting the cells on top, determine what kind of cover is needed to protect the solar cells. Remember they can easily be jarred and broken. They should not be left out in the open in all kinds of weather. Since the radio has a plastic surface, the cells may be cemented directly upon the radio. A small flat type plastic lid was chosen to fit over the exposed cells. Any clear plastic container may be used.

The plastic lid was taken from a 2½ × 4 inch plastic box (Fig. 28-5). Look around the house, you will be surprised how many things come in plastic boxes. If not, you can pick them up any size or shape at hobby and craft stores. This will protect the cells from excessive jarring and any possible accident. Not only will the plastic lid prevent solar cell breakage, but it will keep out the rain. A piece of thin foam was cut to fit inside the plastic box.

Line up the solar cells on the piece of foam. After determining how the cells will mount, remove one at a time and solder a connecting lead to the bottom of each cell. Then lay the cell right back in the line-up. Keep the wire connection as clean as possible. Another reason foam was used for mounting is that these lead connections will settle right down into the foam area and leave the cells level.

After all cells have a bottom connecting wire turned upward, slide the cell around and in line, just so the cells do not touch. Cement the cell to the foam with clear rubber silicone cement. Be careful not to press down too hard. Remember these cells will break on a hard flat surface. The foam mounting area prevents this type of cell damage. Let the solar panel set up overnight or for at least four hours.

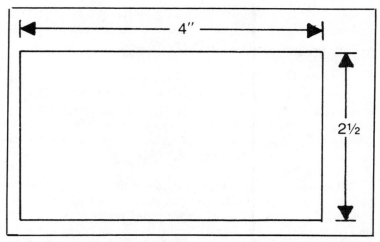

Fig. 28-5. The plastic bar fits over a 2 ½ by 4 inch piece of foam material. The 11 solar cells are mounted on the foam material.

Wiring the Cells

Now the cells are ready to be connected in series. Solder an 8 inch length of hookup wire to cell number one at the top side (−). This wire will connect to the negative battery terminal inside the radio. Take the bottom wire from cell number 1 and lay it upon the center of cell number 2. Cut the hookup wire to the correct length. Scrape back the insulation and tin the exposed wire. Solder this wire to the center bar area of cell number 2. Likewise, connect all cells in the same manner.

Solder a 6 inch hookup wire lead to the bottom side of cell number 11. If a short piece of wire was soldered originally, just connect it to this wire. Tape up or slip a piece of spaghetti insulation over the exposed connection. This positive wire lead will be connected to the positive terminal of the first "C" cell.

Before the solar panel is mounted on the bike radio, check for correct voltage. Connect a vom to the solar panel wires. The red lead of the vom goes to the positive wire (bottom side of cell number 11) and the black lead to the negative connecting wire. In direct sunlight you should have a reading over 5 volts. Under a 100 watt bulb, the solar panel should have an indication of over 3 volts.

Usually, lower voltage than 3 or 5 volts indicates a poorly operating solar panel. Check the solar cells for possible breakage. Look very close since a very fine cracked line across the cell may produce no voltage. Measure the voltage across two cells at a time to determine a poor soldered connection or a broken cell. Repair or remove the damaged cell and install a new one.

Mounting the Solar Panel

Drill two small holes through the plastic top cover after the solar panel checks out. Be very careful. Just go through the top plastic cover. The radio components do not have to be removed to drill the holes, if you are careful. Remove the radio back cover and choose the best place for connected holes to be drilled. Any area along the backside of the radio is okay.

Push the two connecting wires down through the radio top panel. Apply a coat of clear silicone rubber cement to the foam material. Pull out all slack in the wires and firmly press the foam into position. If the top area is curved, be careful not to break the solar cells. With an extreme curved radio top, the foam solar panel may be built up with excessive cement at the ends.

Now, place the plastic cover over the entire solar panel. Line up the solar panel and lid with the back edge of the radio. Cement the plastic cover into place. Use a bead of clear silicone cement around the bottom outside edge of the plastic lid. Take your finger and make a curved seal with the rubber cement. This may take several swipes to achieve a perfect finish. Wipe off all excess cement with cloth or paper towel before it sets up. Let the cement dry for at least four hours.

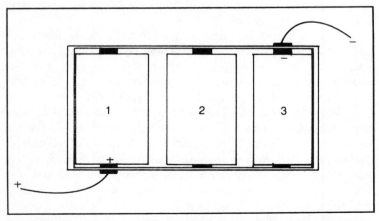

Fig. 28-6. The solar panel is connected to the battery terminals. Connect the positive lead to the positive terminal of the first "C" cell. The negative lead goes to the top negative terminal of cell number 3.

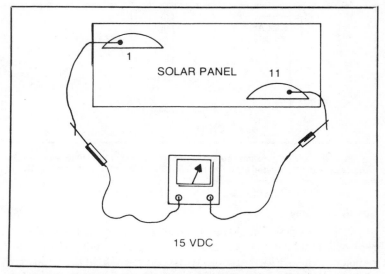

Fig. 28-7. Take voltage measurements across the wires connected to the battery terminals from the solar panel. This voltage should be 3 to 5 Vdc.

Connect the black wire (−) to the negative or spring terminal of "C" cell number 3. The remaining red wire (+) will connect to the positive terminal of "C" cell number 1. Solder these wires to the outside battery connections. The battery compartment of the radio may be removed with two small metal screws. The positive connection of cell number 1 is at the bottom side. Connect the negative wire to the top outside connection of cell number 3 (Fig. 28-6).

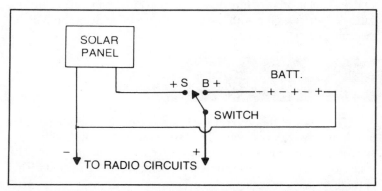

Fig. 28-8. You may be able to operate the radio from the solar panel or batteries by inserting a spst toggle switch. The batteries may be operated on very cloudy days.

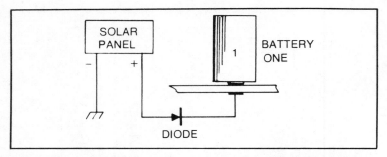

Fig. 28-9. If you desire to charge small radio batteries simply insert a diode in series with the positive battery terminal. Turn the radio off and set the bike out in the sun.

Check Out Time

The bike radio should be checked out before the back panel is replaced. Hold a 100 watt bulb over the solar panel and turn the radio on. Tune in a local station. With this type of light the radio may have weaker volume but should operate. If not, check all connecting wires. Take a voltage measurement across the battery compartment terminals (3 to 5 volts). Check for correct polarity of the solar panel. Re-check those solar panel wires (Fig. 28-7).

The batteries may be removed when operating from the solar panel. If you like, a spst toggle switch may be added to the rear cover for battery or solar power operation (Fig. 28-8). Now, you may operate the batteries on dark days or switch on to solar power. When the batteries get weak, they may be charged by inserting a diode between the batteries and positive connections. The toggle switch is not used with this type of circuit (Fig. 28-9). Just insert the batteries for charging with the radio turned off and set the radio in the sunlight.

PROJECT 29
SOLAR SOLDERLESS BREADBOARD

The heart of this solar breadboard project is a regulated solar power supply with two solderless experimenter sockets. All three separate components are mounted on one piece of masonite. The solar power supply has a rating of 5.6 volts up to 0.225 amps. This voltage source will supply power to most transistors and IC projects. Simply, dream up your own circuit and mount them on the solar solderless breadboard. Power is supplied by the solar cells.

218

Thirteen solar cells wired in series provide over 6 volts to the solderless breadboard. A 5.6 volt zener diode regulates the output voltage for many low voltage projects. Choose solar cells with a current rating of .225 amps or higher. If a larger power supply is needed, choose solar cells with a greater current rating. Remember, the greater the current rating, the larger the solar cells.

Two separate sets of banana jacks are used for current and voltage measurements. An outside vom is used for this purpose. At any time you may be able to measure the voltage applied to your favorite circuit. Just remove a tie wire from the current jacks and you may see at once how much current the circuit is pulling. Not only will you be able to watch the applied current and voltage, but these two tests may indicate a leaky or shorted hookup in your experiment.

Breadboard Construction

The three separate components are mounted on a piece of ¼ inch masonite (Fig. 29-1). Choose the size of the experimenter socket boards to determine how large the mounting board should be. The larger modular IC and pc board solderless socket boards come in 500 or more solderless connections. If two small versions of 270 holes or less are used, the length of the breadboard is less by at least three inches. So choose from two separate modular socket boards before cutting the piece of masonite. These two boards will provide plenty of connections for your electronic projects. See Table 29-1.

Fig. 29-1. Here are the components needed to construct the solar breadboard. Only a drill, saw, and soldering iron are needed to complete this project.

Table 29-1. Parts List for Project 29.

```
 1  — Piece of ¼″ masonite 7×12 inches.
 1  — 4×6 piece of 1 inch foam material.
 2  — Experiment solderless sockets. GC Electronics J4-400 or
      J4-402, Radio Shack, 276-174 or 276-475, John Meshna, Inc,
      SP-65B, Granois Merchandisers, Inc. 12A10320-6 or
      12A10319-B.
 4  — Banana jacks — GC Electronics F2-926, Radio Shack J74-725,
      Gravois Merchandisers 12-01047-6, ETC. TE056, John Meshna Inc.
      SP-258D.
 2  — Male plugs — GC Electronics F2-860, Radio Shack 274-721
      Gravois Merchandisers 12-01042-7, ETCO TE036.
13  — Solar Cells (200 to 250 mA) H & R Inc. TM21K932, John Meshna
      Inc., H-14A, Solar AMP S124, GC Electronics J4-804,
      Poly Pak, Inc. 5307.
 1  — 5.6 1 watt zener diode — GC Electronics J4-1615 (5.1V).
      Radio Shack 276-561 (6.2V) — 5.6V zener diodes found at
      most Radio and TV shops.
Misc. Hook-up wire, nuts and bolts, solder, etc.

Cost — Under $50.00
Construction Time — 6 hours
```

Here two small $2\frac{1}{8} \times 3\frac{5}{8}$ modular solderless socket boards were used. Since these boards are less than ⅜ inch thick, a 1 inch piece of foam material was used to mount the solar cells on. The overall length of the piece of masonite is 12 inches, including the handle cutout area. Check Fig. 29-2 for the masonite mounting board dimensions.

Cut the masonite board outline with a bench or handsaw. This material is easy to cut and form. The handle or hole area of the breadboard may be cut out with a saber or coping saw. If a router is handy, taper out the handle area. Lay the face-side of the masonite down on the saw area so the finished top edge will be clean cut and not ragged. Sand down all edges.

Drill all holes to mount the two experimenter solderless sockets through the masonite. Each modular socket has four mounting holes. Lay the boards on the piece of masonite. These boards may be mounted side-by-side for easy wire hookup or spread two inches apart for a balanced appearance. Ream out or countersink the holes at the back of the piece of masonite. Since the bottom side must lie flat, use flat headed screws with nuts at the top side of the modular socket boards.

The piece of masonite should be sprayed with paint or a clear finish. A couple of coats of paint or clear finish will give a pleasing appearance. Use either clear lacquer, varnish, or a protective coating spray. Lightly sand between coats. Use a tack cloth to remove any particles. Let the masonite mounting board dry for a couple of days and in the meantime build the solar panel.

Solar Cell Hookup

A commercial type solar panel may be used with this project or you may want to construct your own. Select a 1 inch thick piece of foam material. The solar cells and banana jacks will be mounted on this piece of foam material. Cut the piece of foam 3 × 6 inches. This foam material may be cut with a pocketknife or any type of a saw. Drill all holes for the current and voltage measurement jacks. A pocketknife or awl may be used instead of a drill. Use the pocketknife blade to countersink the banana jacks mounting nuts up inside the bottom foam area.

After all holes are cut and drilled, the solar cells may be mounted. Cut a ¼ inch length of hookup wire for each solar cell. Tin one end of the piece of wire. Solder a wire to the bottom side of the 12 solar cells (Fig. 29-3). The last cell should have a 6 inch length of hookup wire soldered to the bottom. Keep the soldering joint clean and without heavy soldered areas, although this excess solder will

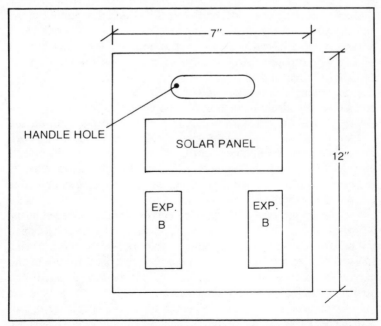

Fig. 29-2. This is the layout for the piece of masonite. Saw and drill all holes before sanding.

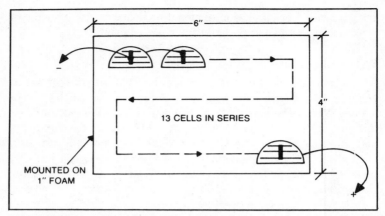

Fig. 29-3. Connect the thirteen solar cells in series to acquire the 5.6 regulated voltage. Use one quarter or half round (.225 to .250 amp) cells for the project.

push down into the soft foam material. Now, the cells are ready for mounting. Lay the cells out in a straight line.

Usually 225 or 250 mA solar cells are one-fourth or one-half sections of a round solar cell. Now, place a coat of clear silicone rubber cement on the bottom of each cell. Keep the cement in the center of the cell. Press the cell down upon the foam material. Don't press too hard or you may break the cell. Wipe off any cement excess with a paper towel and the edge of a pocketknife. Let the cells set up overnight.

Start with cell number one when the cells are dry and can be connected. Bend the insulated wire down to the center bar of the solar cell and cut off with a pair of side cutters. Scrape off the insulation and tin the hookup wire. Solder and connect each cell until all cells are connected in series. Now, solder a 6 inch connecting wire to the top of cell number 1. This wire of cell number one, is the negative terminal and the bottom connecting wire of cell 13 is the positive terminal.

Connect the two wires to each corresponding voltage banana jack (Fig. 29-4). The positive terminal should go to the red jack and the negative terminal wire should be soldered to the black banana jack. Now, run a wire from the positive (red) jack to one of the current jacks. The other current jack will be connected later to the positive terminal of each modular board.

Before the solar panel is cemented to the masonite breadboard, check the voltage output across the voltage jacks. Insert the vom into the voltage test jacks. The common or black lead should be inserted into the black jack while the red lead of the vom is inserted into the red jack. Under a reading lamp, you should have around 4 volts dc. In direct sunlight, the voltage should be greater than 6 volts.

When the voltage is a great deal lower than normal, suspect a bad soldered connection or cell. Measure the voltage across two cells at the same time. Now, compare these measurements with the next two cells. You should have slightly under 1 volt when two cells are normal and are connected in series. Repair the connection or discard the defective solar cell.

Mounting the Components

Before mounting the solar panel to the masonite board, solder the zener diode across the two voltage banana jacks. The diode may be inserted later across x and y of the solderless sockets (Fig. 29-5). Be sure the white line (+) terminal of the diode is connected to the red jack terminal. Cut out some of the foam to place the diode inside, out of the way. Now, bolt the two experimenter boards into position. Secure the nuts on top so the bottom side of the masonite remains smooth and level. Place a dab of cement over each nut so they will not loosen up.

The modular board wiring may be run on the outside and plugged into each board. If a more permanent connection is desired, solder the connecting wires to the terminal strips underneath. Peel

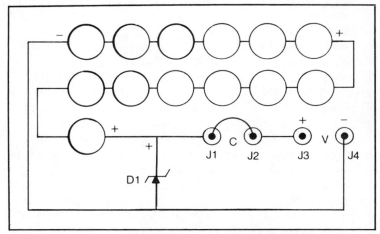

Fig. 29-4. Here is the wiring circuit of the solderless breadboard. Take a voltage measurement (5 V) at the voltage jacks with the solar panel under a 100 watt bulb.

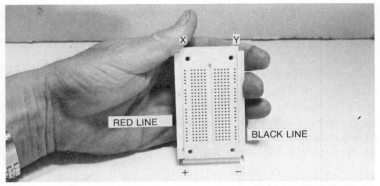

Fig. 29-5. The zener diode is connected across the voltage output terminals. Always, observe correct polarity. Connect the red and black wires to the x and y connections of each experimenter modular board.

back the plastic paper until the socket bars are exposed. Now, drill a very small hole through the plastic above the letters x and y. This will allow the board to mount flush on the piece of masonite.

Wiring

Run a wire from the red voltage terminal jack to the letter "x". Push the hookup wire down through the plastic. Cut the hookup wire the right length by temporarily laying the modular socket into position. Pull all the wire slack down through the small hole. Now, solder this lead to the "x" metal bar. Likewise, do the same with the "y" terminal. Connect a black wire to the negative jack and solder it to the "y" terminal.

Insert a small wire at the x and y socket connections. Check the voltage at these two points. If the voltage is normal (5.6 V) secure the two modular boards to the piece of masonite. You may want to just insert the voltage source wires to the x and y sockets of each board after they are mounted. Although, the job is more permanent with soldering the connections, wire insertion works as well.

When completed, the solar solderless breadboard may be placed into action. Remove one end of the current plug so voltage will not be applied to the circuit. Construct your favorite experiment by inserting the components in the solderless socket. Set the vom current range to 100 mA and insert the meter prodes. As your circuit comes to life, you may see directly how much current it's pulling. If the circuit is wired up wrong, the meter will hit the peg, indicating a short or leaky circuit.

The solar solderless breadboard may be operated in the sunlight or under a large incandescent bulb. When the project is to be operated under a fluorescent bench-light you should add more solar cells to acquire the correct voltage. You will find larger round cells work best under fluorescent lights. One thing for sure, you don't have to worry about batteries going down or leaking out while using the solar breadboard.

PROJECT 30
PORTABLE SOLAR POWERED CB

Just add this small power supply and you may operate that small CB unit with solar power. Let the solar power supply ride piggy-back or on top of the CB unit. You can talk to your heart's content and not ever worry about wearing out those batteries. If you like, the solar power can be connected to help charge small batteries, in or out of operation.

Adding the Commercial Unit

There are several small commercial solar power panels on the market which may be attached or connected to the portable CB unit (Fig. 30-1). Most portable CB walkie-talkies operate with 4.5, 9, 12, or 15 volt batteries. In the 9 volt unit a single 9 V battery is found.

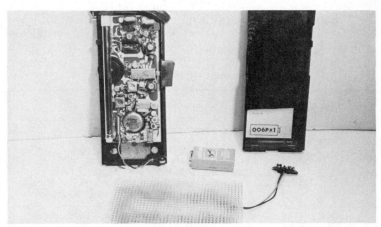

Fig. 30-1. Select a 9 volt commercial solar panel for your small walkie-talkie unit. Attach Velcro to the back of the panel and CB unit for easy removal.

While in others, 6, 8 and 10, 1.5 volt batteries power the walkie-talkies. So choose the solar panel, with the correct voltage. See Table 30-1.

The solar panel may be constructed from units operating with less than 1 watt. Portable CB units above that power rating may pull too much current for practical purposes. Although, the solar panel may be used to charge the small batteries when not being in use. In fact, many of the larger powered units operate with 8 or 10 rechargeable batteries.

Roll Your Own 9 Volt Solar Supply

First, determine how much current the CB walkie-talkie needs to operate. Slip a piece of cardboard between the positive terminal and battery. With units using a small 9 volt battery, remove the positive clips and turn the battery sideways (Fig. 30-2). Connect the vom between the battery and contact to determine how much current is drawn. Set the vom to the 30 mA range. Now, turn the CB on and speak into the microphone. Take the average current reading to determine what solar cells are needed. You will note the receiver may pull around 10 mA in receive and up to 25 mA in the transmit position.

For instance, if the meter reads 100 mA of current, select cells of the 100 to 250 mA range. It's always best to select larger solar cells than are actually required. Of course, the greater the current, the larger the cells. The larger the cells, the greater area to mount them. Since most CB units are 3 inches wide and from 6 to 10 inches high, the solar panel will have to stick about 5 inches above the top of the case for normal operation.

Table 30-1. Parts List for Project 30.

```
Commercial solar panel — 9 volt solar amp #S100
Universal 50 mA solar panel — solar amp #S116, Poly-Pak, Inc.
          #6413, Edmund Scientific #42,746.
Solar panels:
          20 solar cells 100 mA — Edmund Scientific #42,268,
          Solar Amp #S122, John H. Meshna, Inc., #H14A,
          H & R Inc., #TM21K666, Poly-Paks, Inc., #S306 or S538.
          20 Solar Cells 50 mA — Poly-Paks, Inc., #5538, H & R, Inc.
          TM20K187, John Meshna, Inc. #H-11, Solar Amp #S122

Misc.   9 volt socket with leads, ⅛ inch plastic panel,
          6 inches Velcro.
Construction Time — 5 hrs.
Cost — $19.95 to $29.95
```

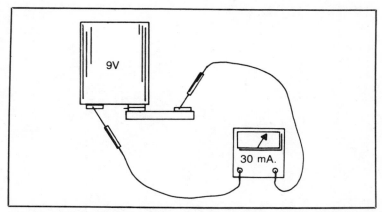

Fig. 30-2. Connect the vom to the CB to determine the operating current. With units using a 9 volt battery, remove the positive clip and turn the battery sideways.

A 100 mA 9 volt power supply requires twenty 100 mA or greater solar cells. Twenty 100 to 150 mA solar cells (¼ round) will fit nicely in a 4× 7 inch panel. The solar panel will protrude ½ inch on each side of the walkie-talkie case and 4 inches above. While using 225 or 250 mA (¼″) solar cells, the panel may be 3 × 8 inches. Of course, these larger cells may be cut (Fig. 30-3); which requires a total of ten 225 mA cells cut in half.

After selecting the required solar cells, place them on a piece of paper. Three of these 100 mA cells may be placed side-by-side, within the 4 inch width area (Fig. 30-4). Select the correct width, not more than ½ inch wider than the body of the walkie-talkie. One 225 or 250 mA cells may be mounted in the three inch width since they are cut in half.

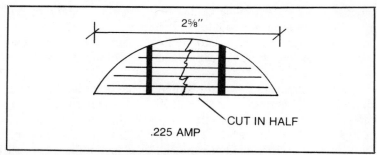

Fig. 30-3. Larger solar cells may be cut in half for easy mounting. Always choose half round or crescent cells for this purpose.

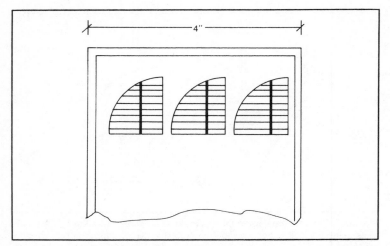

Fig. 30-4. Line up the solar cells on a piece of paper. Three of these 100 mA cells may be placed side-by-side, within the 4 inch width area.

Cut two pieces of plastic for the 4 × 7 inch panel. The back panel should be ⅛ inch thick with a thinner top panel. Cut a piece of ⅛ inch plastic to fit around the edges of the solar panel. This small ¼ inch piece will serve as a spacer between the top and bottom panel (Fig. 30-5). The panels and spacer may be bonded together with regular plastic cement.

For the 3 × 8 inch solar panel, cut two plastic pieces, 3 × 8 inches overall. The front piece may be thinner than the ⅛ inch back panel. Again the two plastic pieces are spaced with a ¼ × ⅛ inch piece of plastic. This plastic spacer will go around the outside edge. Of course, the plastic spacer on the top panel is not bonded to the bottom panel until the solar panel has been tested.

The solar cells are mounted directly upon the bottom plastic piece. First, lay the cells out in line so they will fit in the required area. Lift one cell up and solder a ¼ inch lead to the back side of the cell. Then lay the cell back into the line. Solder a lead to all cells except the last one. Here, solder a six inch lead of wire to connect to the 9 volt socket.

Now, cement all cells into position. Place a large dab of silicone cement on the bottom of each cell. Press the cell down, but be careful. Remember the connecting wire and solder joint is under the cell preventing a tight level fit. This may hold the cell up from the hard piece of plastic. Just keep the cell level and in line. After all cells are mounted, let the panel set up for at least 4 hours.

Start at cell number 1 and connect the piece of wire from the bottom side to the center bar terminal of each cell. All 20 cells are wired in series for an efficient 9 volt operating solar panel. Cut the wire so it reaches the center bar of each cell. Clean back the insulation. Quickly solder the hookup wire to the cell. Be careful not to melt down the bonding cement. After a few attempts, it's easy to solder each one. Connect an 8 inch length of hookup wire to the top of cell number 1.

Bring the two remaining connecting wires down to the bottom of the cell. To hold the wires in place, apply the iron to the wire and let it sink into the plastic, or apply a dab of silicone cement to hold the wires in place. If a regular 9 volt battery is used, connect a 9 volt battery socket (with leads) to the bottom of the solar panel. Place rubber cement over the bare soldered connections. Apply cement to hold the flexible cord and plug from pulling out of the solar panel.

Make a voltage check before placing the top plastic cover over the solar cells. Place the panel in the sunlight and you should have a reading of 10 volts. Under a 100 watt bulb, the voltage reading should be 7 to 9 volts. If the voltage reading is lower than these

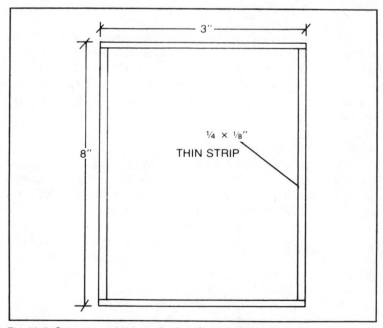

Fig. 30-5. Cut a piece of ⅛ inch plastic to fit around the edges of the solar panel. This small ¼ inch piece will serve as a spacer between the top and bottom panel.

readings, suspect a poor soldered connection or cell. It's best to take a voltage reading across two cells with a 100 watt light above the solar panel. Now, check these readings and compare across the next two cells until the poor connection is found. A defective or broken cell will definitely have a lower voltage measurement.

Mount the side plastic spacers and top panel. Use regular plastic cement for bonding plastic to plastic. This liquid cement may be obtained from hobby and other retail stores. Irregular gaps in the plastic ends may be filled with clear silicone rubber cement.

Roll Your Own Solar Power Panel

For small walkie-talkies (less than 25 mA), small solar cells may be used. Check the current with the vom under operating conditions. You may find when receiving, the CB pulls only 10 mA of current, but, when transmitting, the unit operates between 20 and 25 milliamperes. Here small solar cells may be used to power the small CB unit.

With a 25 mA current operation, choose 50 mA solar cells. You will find the mounting space is a lot smaller and will fit nicely on the rear of the CB unit. For a 9 volt panel, you will need 20 cells. A 6 volt solar panel requires 15 solar cells.

Since these quarter or crescent cells are physically small, a 2½ - × 6 inch solar panel is needed. Cut two pieces of plastic the same size. The back plastic panel should be at least ⅛ inch thick. Mount and wire all cells in series as in the 3 × 8 or 4 × 7 solar panels. Connect and test for correct voltage before applying the sides and top cover.

Mounting the Solar Panel

The solar panel may be cemented to the back of the walkie-talkie with silicone rubber cement. For a temporary or easy removal of the solar panel, use a product called *Velcro*. When two pieces are placed together, they are bonded together. The two may be separated by pulling the units apart. Velcro can be picked up at most sewing centers or department stores. The product comes in circles, squares and strips. A 1½ inch strip was used in mounting the solar panels.

Cement a 1½ inch strip, 4 inches long of Velcro on the back of the CB unit and another the same dimensions on the back of the solar panel (Fig. 30-6). Let the cement dry overnight. Now, the two units may be stuck together. Just pull the solar panel from the CB for

Fig. 30-6. Cement a 1½ inch strip, 4 inches long of Velcro on the back side of the CB. Cement the other matching piece to the back of the solar panel.

easy removal. You may want to add Velcro to several other solar projects where this same 9 volt solar panel may be used.

Connecting the Solar Panel

The solar panel connections may be snapped together with regular 9 volt battery connections. Regular plug and jack sets may be used to connect the solar power to the CB unit. Mount the female jack on the CB with the male plug attached to the panel cord. The solar panel may be wired directly into the battery compartment.

Remove the batteries and check the wires going into the CB. In most cases you will find a red and black wire. Connect the positive lead from the solar panel to the red lead. Wire the negative lead to the black wire connection. Double check the wiring. Notice if the positive end of the battery goes directly to the red wire. If you desire to have the solar panel charge the batteries when not in use, insert a diode in series with the red lead of the solar panel. This will prevent discharging the batteries into the solar panel.

Testing

After the solar cell is connected to the CB unit, give it a test. Take it outdoors and point the solar panel towards the sun. Turn the switch on and listen for a rush from the speaker. Make sure the batteries are removed. Now call your neighbor or fellow CB'er.

The CB unit may be checked out inside the house under a 100 watt bulb. Place the bulb near the light and listen for a rush in the speaker. You may note the volume of the CB unit is a little lower than in the sunlight. A 9 volt solar panel may not register more than 7 volts under a 100 watt bulb. But, most walkie-talkies will receive, indicating the solar panel is working. Measure the voltage at the solar connections with the CB unit working under load.

Chapter 5

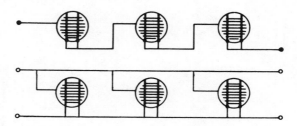

Three Solar Projects above $50

**PROJECT 31
DELUXE SOLAR PANEL**

Here is a medium-priced solar panel that you can make your-self. This panel may be used to charge the car, boat, or solar electric batteries. You may also use the solar panel for the recreational vehicle, to operate the cassette player, or a small solar television receiver. The panel may be used to charge the small batteries in those three-way portable B&W TVs. Of course, this solar panel may be employed to furnish power to most other electronic projects in this book.

Panel Construction

The 36 solar cells are mounted in two rows making the panel dimensions 10 × 30 inches (Fig. 31-1). These solar cells are mounted upon a ¼″ piece of masonite material. Cut the piece of masonite 10 × 30 inches. Sand down all corners with a portable sander. Use the rough or backside to mount the cells on. One quarter inch plastic may be used to mount the solar cells on. See Table 31-1.

Cut two masonite strips 1″ wide to go down the sides. The same type of material will go into the two ends. Lay the strips on the masonite panel. Now, drill a ¼″ hole in each corner for mounting.

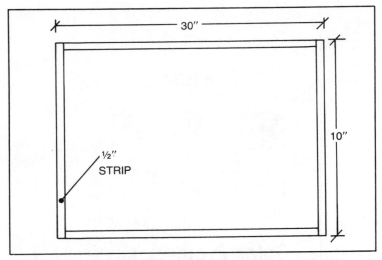

Fig. 31-1. Cut a piece of ¼ inch masonite 10 × 30 inches long. Sand down all rough edges. A one inch strip is fastened around the outside edges.

Draw a line inside of the strip area. Then remove the strips as they will be glued to the bottom panel after the cells are mounted and wired.

Choosing the Cells

Choose ½ cells of either 2¼ or 3 inch round solar cells. These cells come in 400, 500 and 600 mA sizes. The cost of each cell may vary from $4.00 to $9.95 depending upon where they are purchased. Sometimes manufacturer cutoffs and overrun cells may be used, lowering the purchase price.

The 36 cells are wired in series for a 12 volt 400 to 600 mA output (Fig. 31-2). The row of cell number 18 is center tapped so

Table 31-1. Parts List for Project 31.

36 Solar cells 400, 500 or 600 mA half round types
    Solar Amp #S123, Poly-Pak #5312, Edmund Scientific Co. #42,728
    Manufacturers cut off and over run cells may be used here
    for a lower priced solar panel.
1 — 10×30 inch piece of ¼ inch masonite material
1 — 10×30 inch piece of ⅛ inch plastic
8 — soldering eyelets or terminal strip
Misc. Hookup wire or cell tabs, nuts and bolts, silicone
    rubber cement, sealing material, etc.
Cost — Solar cells cost $150 to $225
Construction time — 10 hrs or more

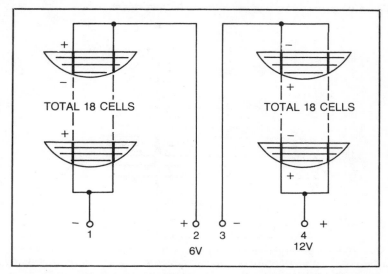

Fig. 31-2. The 36 cells are wired in series for the 12 volt output. Connect the cells in a parallel-series arrangement for 6 volt operation.

that a 6 volt output operates at 1 to 1.2 amps. The two rows of solar cells are wired in a series, parallel arrangement for 6 volt operation. Of course, the operating current will double when both rows of cells are wired in parallel.

The solar cells may be connected with tabs or tie wires. You will find some cells with a single or double line down the middle of the cell. Double grid or lines should be connected to the next cell. When flat wire or tabs are used, first solder them to the bottom of each cell. The same applies to single grid line cells. This will keep the cells in a straight line.

Cut 36 pieces of flexible wire one inch long to connect the cells together. Flat nickel wire is ideal. Some cells come with flat tap connections while others may have black and red connecting wires. Usually, the lower priced cells do not have any connecting wires or tabs for connecting together. When no wires are attached, solder a flexible wire to the bottom side right below the grid line. Small flat copper shield material may be used for connecting the cells together. With this method the cells may be mounted one at a time.

With one-half round cells without any attached tabs or wires, solder a one inch wire on the bottom side in line with the top grid lines. If two grid lines are found on top, two separate wires should be soldered to the bottom side (Fig. 31-3). Solder the one inch lead

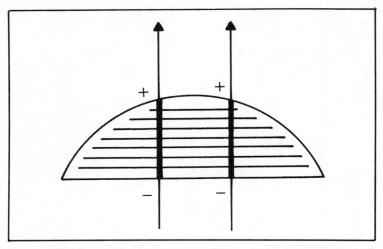

Fig. 31-3. When two grid wires are found on the top of the solar cells, solder both grid connections to the next cell. You will have two connecting wires or strips to each cell.

to the first 18 solar cells. The two top cells of each row should have a long enough wire to be able to connect them together.

Place the cells on the masonite panel so they are about ½ inch from the outside line. Draw a line down through the two rows where the cells will mount. Also draw a line ½ inch from the outside line so the edge of the cells will start at this point. Try to keep the cells square and in line down the whole panel.

The eighteen solar cells may be wired in series before mounting or mounted separately. When mounted separately apply clear rubber silicone cement to the back of each cell. Apply enough rubber cement so the cell will lie flat against the masonite panel. Lay the cell in line and pull up the connecting wire that will solder to the next cell. These cells can be moved and lined up before the rubber cement sets up. Be very careful not to push down on the cell and crack the top surface. Mount all 36 cells in the same manner. Let the cells set up overnight. Now, connect the loose wire to the next cell. You are now connecting each cell in series with the next one. The connecting wires should be soldered to the top grid bar of the next cell. Make a good soldering connection but do not apply too much heat to loosen up the cell mounting cement. Check the cell voltage before going any further.

Measure the voltage across the 18 cells with the vom. You should have from 6.5 to 8 volts under strong sunlight. Check the wiring and cells when the voltage is under 6 volts. You may have a poor soldered connection or a defective solar cell. Check the cell surface for cracked areas. Remove the damaged cell and replace it. Measure the voltage across both rows of solar cells. Make sure each row of cells measure from 6 to 8 volts in strong sunlight.

Before connecting the cells in a series-parallel arrangement apply rubber cement around the outside edge of each cell. This will hold the cell in place and keep out any moisture that may form within the panel. Connect the two sections of cells together. Now, all 36 cells are in series. Tap off a lead between the two sections for a 6 volt output. From left to right mark off each terminal as 1, 2, 3, or 4. Terminal 1 is the negative terminal for both 6 and 12 volt sources. For 12 volt operation, connect 1 and 3, and 2 and 4 for a series-parallel output (Fig. 31-4).

The voltage tie points may be connected to the project with flexible wire such as rubber ac cord. Another method frequently used is with soldering eyelets with terminal screws and nuts (Fig. 31-5). Three 8/32 bolts and nuts are fed through holes of the masonite board. Use a soldering eyelet for each wire going to the solar cells. Apply washer and eyelet at the top of each terminal for easy feed off from the solar cells. Another method is to use a four screw terminal strip mounted on the bottom or top panel (Fig. 31-6).

Finishing the Project

Before sealing up the panel, make sure the correct voltage is obtained at the 6 and 12 volt terminals. The solar panel may be

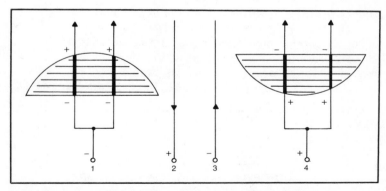

Fig. 31-4. For a 12 volt operation, connect terminals 2 and 3 together. The 12 volt output terminals are now 1 and 4. For a 6 volt series-parallel arrangement, connect terminals 1 and 3, and 2 and 4 together. Terminals 2 and 4 are the positive output terminals.

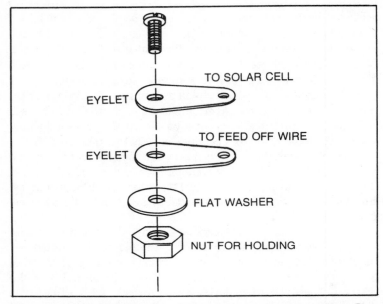

Fig. 31-5. Use soldering eyelets or dog ears for easy power connections. Place one eyelet for the solar cell connection and one for the voltage output terminal.

sealed with several coats of liquid plastic. Place at least two coats on the solar panel. Let each coat dry thoroughly before brushing on the next one. A piece of clear or indented plastic may be cut to cover the front of the entire solar panel. Either bolt or use pop rivets to hold the masonite and plastic panel together. Seal all ends with clear rubber silicone cement, liquid plastic, or bonding material.

Another method is to use silicone rubber with a hardening agent. Be careful, the rubber may set up within 15 to 30 minutes. Heavily brush the rubber over the entire solar panel. All cells are embedded in the rubber mixture. The whole solar panel may be enclosed in liquid rosin plastic. Check your local hardware store and

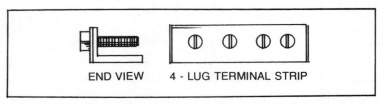

Fig. 31-6. A four screw terminal strip may be mounted at the bottom edge of the panel for a voltage output terminal. Number the voltage terminals 1, 2, 3, and 4.

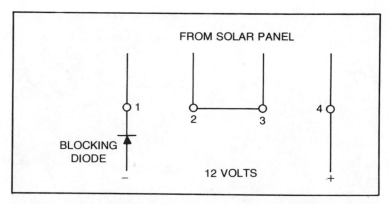

Fig. 31-7. Insert a fixed silicon diode in the negative leg of terminal 1. The cathode or positive terminal should connect to the solar cell.

solar outlets for these products. Be cautious of fire hazards and chemical fumes. Mix the liquids outdoors.

Before mounting the solar panel in a permanent place, check for correct 6 and 12 volt outputs. The panel should be mounted in the sun, but out of the way. Rubber ac cord may be used to connect the solar panel to a terminal block inside. A flexible cable with two large clips may be used to connect the charging battery.

Place a 1 or 2.5 amp silicon diode in series with the negative terminal 1 for battery discharge protection. The positive terminal of the diode will go to terminal 1 (Fig. 31-7). With the diode connected in the negative lead this provides protection for both 6 and 12 volt operation.

PROJECT 32
COMMERCIAL SOLAR PANEL KIT

There are very few solar panel kits on the market for those experimenter's interested in assembling their own. In most cases you have to build the solar panel from scratch. Here is a sturdy, well-constructed 12 volt 750 mA solar panel kit. The sides are cut from aluminum bars with a fairly thick aluminum backside. There are a total of 32 three-inch solar cells connected in series which are embedded in a weatherproof silicone rubber or liquid resin.

Step-by-step instruction booklets with drawings are very easy to follow. Besides the step procedure and drawings, several photos are added to point out how to assemble the solar panel. A few added

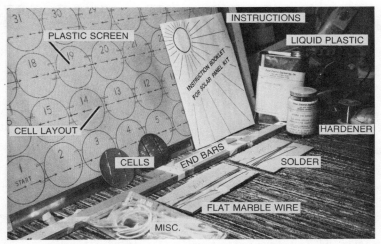

Fig. 32-1. The solar panel kit consists of four aluminum side bars and a sheet of aluminum for the bed panel. Self-threading screws hold the bed and bars together. The kit includes everything to assemble the solar panel, right down to a small brush and stirring stick.

tips are given as we put the panel together. Here is a practical solar panel that should last for years.

The solar panel kit consists of four aluminum side bars and a sheet of aluminum for the bed panel (Fig. 32-1). Self-threading screws hold the bed and bars together. A sheet of sandpaper is included to rough up the metal bed or the silicone rubber will not stick to the metal bed area. Thirty-two 3-inch solar cells provide the 12 volt output source. See Table 32-1.

Anyone can connect the solar cells by using the pictorial guide. A piece of insulating screen insulates the solar cells from the metal bed. After all cells are in place a can of clear liquid resin is mixed with a bottle of hardener providing a weatherproof solar cell panel. Everything needed for building the solar panel is included in the kit,

Table 32-1. Parts List for Project 32.

Solar Cells — 3 or 4 inch solar cell at 1 amp or greater (listed in alphabetical order.)
Edmund Scientific 101 E. Gloucester Pike, Banington, NJ 08007
H & R Inc., 401 E. Erie Ave., Philadelphia, PA 19134
John J. Meshna, Inc., P.O. Box 62, E. Lynn, MA 01904
Poly-Paks Inc., P.O. Box 942, So. Lynnfield, MA 01940
Radio Shack (at locally owned stores)
Silicon Sensors Inc., Hwy 18 East, Dodgeville, WI 53533
Solar Amp. P.O. Box 27885, Denver, CO 80227
Miscellaneous: Sheet aluminum, masonite, hookup wire, terminal
take-off strip (Radio Shack, 274-315), glue, liquid plastic, etc.

right down to a brush and stirring stick. Even a diode is included to prevent current from the charged battery from discharging through the solar panel.

Step I: Cutting Wire for the Panel

Cut the flat nickel wire to connect the cells together and for the positive and negative feed-off terminals. Two pieces of flat wire are cut 6 inches long which will be soldered to the top and bottom sides of solar cells 1 and 32. Now, cut off 35 pieces of flat wire one inch long, to connect the cells together (Fig. 32-2). When double line cells are used cut off 70 pieces, instead. You will have two connecting wires from each cell.

Use a pair of tin snips or a regular pair of scissors to cut the flat nickel wire. The wire is very easy to cut. To speed up the process when 70 pieces are needed, clamp a piece of board on one inch of the ruler. Now, just butt the end of the flat wire to the board and quickly snip off the end. Make sure all kinks are out of the flat wire before cutting. Pull the flat wire over the edge of a board to remove all bent or kinked areas.

Step II: Preparing the Flat Connecting Wires

Turn the flat nickel wire before soldering to the solar cells. You may do all 35 or 70 pieces just before you solder the connecting wire to the cells. The wire tins very easily.

Fig. 32-2. Cut 35 pieces of flat nickel wire to connect the solar cells together. You will need 70 cut pieces if double line cells are used in the project. A regular pair of scissors or tin snips may be used to cut the thin wires into pieces.

Step III: Soldering Wires to the Cells

Now solder the flat nickel wire to the top side of each cell. Solder two connecting wires for double line solar cells (Fig. 32-3). Be very careful. Do not apply too much pressure on the cell area. Lay the cell flat against a board or book so it will not be easily broken. Keep the flat connecting wires straight with the cell.

When a soldering gun is used, do not leave it on the connection too long or you may damage the cell. Use a low wattage iron. Be careful not to lift the grid solder from the cell. Let the solder run between flat connection and the cell. Take your time when soldering the connecting wire. A cold soldering joint may result in poor output voltage.

Step IV: Sanding the Aluminum Sheet

The heavy aluminum sheet forming the bed for the solar panel should have the topside sanded (Fig. 32-4). This provides a matt finish so the liquid silicone rubber or resin will cling to the metal sheet. Sand down only the topside. Do this before adding the heavy metal side rails. If not, you won't be able to sand properly in the corner areas.

Step V: Bed Panel Construction

Before attempting to place the metal end bars to the bottom panel clean off all sticky tape residue with alcohol. These end bars come taped together for easy packing. First, mount the end bars,

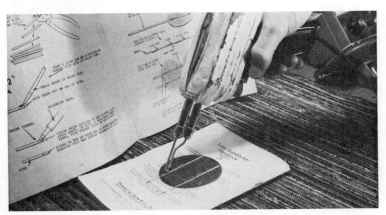

Fig. 32-3. Solder the flat nickel wire to the top side of each cell. Two connecting wires are needed for double line cells. Be careful not to break the cells. Make a good soldered connection.

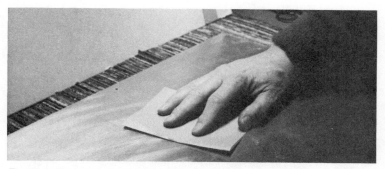

Fig. 32-4. Sand down the topside of the metal aluminum bed sheet. This provides a matt finish so the liquid silicone rubber will stick to the metal sheet. Sand only one side of the metal sheet.

make sure the bars are flush with the outside edge of the metal sheet. You may have to turn end bars around to line up the holes for a flush appearance. Since all bars may be tapped into position start the screws in the end pieces (Fig. 32-5).

Step VI: Alignment of Side Bars

Now mount the side bars. Lay the bars down flat on the workbench or table and screw the bottom metal panel to them. Check for flush sides with the holes lined up. You may have to turn the side bars around. Make sure the bars fit tightly against one another. Tap the bars in place, if needed. Then tighten up all metal screws. Be careful, these small screws may strip the bar holes if tightened too tight. Just snug up the metal screws for a tight fit. You

Fig. 32-5. Just start the metal screws into the end pieces, since all bars may be tapped into position. Make sure the bars are flush with the edge of the metal sheet of aluminum. Tap the metal sides in place with a hammer.

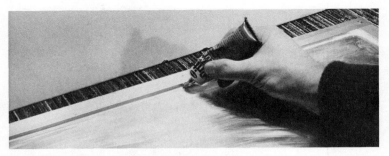

Fig. 32-6. To keep the liquid silicone rubber from running out of the bar seams, seal the seam with a tube of glue. Let the glue set up overnight.

will find that the side bars and bottom panel makes a very sturdy solar panel base.

Step VII: Sealing the Panel Frame

Use the small tube of glue, included in the kit, for sealing the edges between side bars and metal plate (Fig. 32-6). Place a bead of glue in the groove. Sometimes the glue doesn't want to stick, so you may have to go over the area. Let the glue set up for about four hours or overnight. Inspect the glued edges. When some areas of the groove seem to be open apply another layer of glue in these areas. You may have a lot of glue left over after these areas are touched up.

Step VIII: Mounting the Voltage Take-Off Terminals

Two plastic insulated washers are provided to insulate the metal voltage takeoff terminals from the metal panel bed area. Each washer has a lip that will fit down inside the panel holes. Place the nuts on the outside of the metal panel (Fig. 32-7). A dog ear soldered

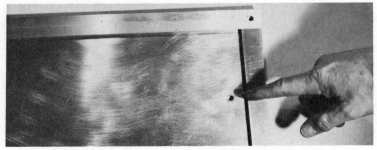

Fig. 32-7. The dog-ear take off terminals are insulated from the metal bed with plastic space washers. Each washer has a lip that will fit down inside the panel holes. Place the dog-ear terminal towards cells number 1 and 32.

Fig. 32-8. Place a dog-ear terminal on top over the bottom side of the metal panel. To keep the nut from loosening up on the back of the panel, place silicone rubber glue or cement over the area.

terminal is provided to solder the flat nickel wire to these posts from solar cells numbers 1 and 32. Likewise, the voltage takeoff dog-ears will be attached to the back to connect the panel wires (Fig. 32-8). Make sure these connections are good and tight.

Step IX: Using the Guide Card

A handy guide card is included in the kit for easy mounting of the solar cells. Just lay the cells in the circled areas and solder up the back side of each cell. Each cell is marked from 1 to 32 (Fig. 32-9). Of course, before the cells can be mounted each cell must

Fig. 32-9. A handy guide card is included to line up the solar cells. Just lay the cells within the circular areas and solder the back of each cell. These cells are marked from 1 through 32.

have the connecting wires soldered to the front. Lay the cells face down on the guide card.

Start with cell number 1. Make sure the takeoff grid wires are connected to the front side and turned towards the terminal lug. These wires will go to the negative terminal post. Line up cell number 2. Place the flat connecting wires on the back of cell number 1. Solder these two grid wires to the metal soldered area. Keep the wires flat and straight. Keep the cells within the circled area.

Lay out and solder each cell as you proceed. Solar cells 8, 16, 17 must have the flat wires turned up at right angles to connect to the next row of cells. Use a pencil or pocketknife to hold the flat wire in place while making a flowing type soldered connection. Be very careful not to crack or break one of the large cells. Turn the connecting wires of cell 32 towards the terminal connection. These wires will be soldered to the positive terminal post.

Step X: Completing the Cell Assembly

After all cells are soldered together, apply scotch tape to the back of the cells. Place the tape up and down across the set of four solar cells (Fig. 32-10). This will hold the cell columns together as the cells are moved from the guide card to the bed panel. Be very careful when handling the column of cells.

Step XI: Installing Screen and Cells in Panel

Place the plastic screen inside the metal panel bed before the

Fig. 32-10. After all cells are soldered together, apply scotch tape to the back of the cells. Place the tape up and down across four solar cells. This will hold the cells together as they are moved from the guide card to the bed panel.

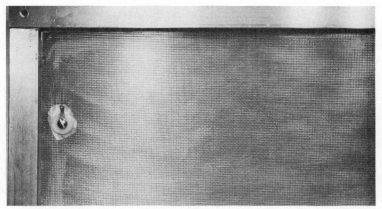

Fig. 32-11. Lay the plaster screw down on the metal bed area. Cut two holes in the plastic screen to let the dog-ear terminals stick through for easy connections.

cells are installed. Cut two holes in the plastic screen to let the dog-ear terminals slide through for easy connection (Fig. 32-11). Center the cells within the solar panel. Solder the takeoff wires to each terminal lug. Check for well soldered connections. The cell column may be placed on a piece of heavy cardboard with grid lines upward. Now the cell column may be pushed off the cardboard onto the plastic screen.

You may want to check the voltage output of the solar panel before the cells are embedded in the liquid resin (Fig. 32-12).

Fig. 32-12. You may want to check the voltage output under sunlight before the cells are embedded in silicone rubber. Connect the vom across the terminal lugs. A very low voltage measurement may indicate a cracked cell or poor connection.

Connect the red probe to the positive terminal and the black probe to the negative terminal of the solar cells. Under direct sunlight you should measure between 13 and 14.5 volts. Here under photo light while the picture is taken the voltage was 17.08 volts. When voltage is lower than 13 volts suspect a broken cell connection or a defective cell. Go over each soldered connection. Check each solar cell for a fine crack or for breakage.

Step XII: Mixing the Silicone Rubber

The panel assembly must be completely finished before mixing the liquid resin. You will find a can of liquid resin with a small bottle of hardener liquid. Pour the liquid in a large coffee can for easy mixing. The complete contents of the hardener is poured into the can of liquid resin. Remember the liquid resin will set up in 15 minutes at 70 degrees. So, stir the contents thoroughly before embedding the cells. Keep the mixture away from open fires.

Step XIII: Embedding the Cells

After the liquid resin has been thoroughly mixed with the hardener agent, pour the entire contents over the solar cells. Make sure the panel is level before pouring the liquid. Cover all cells with the aid of a small brush (Fig. 32-13). Double check all cells. Take the brush and press lightly on each cell to work air bubbles out from under the cells. Let the solar panel set up overnight.

This is how the finished solar panel will look after it is completed (Fig. 32-14). Now take a voltage output test under direct

Fig. 32-13. Make sure the panel is level before pouring the liquid rubber. Cover all cells using the small brush. Lightly push down on cells to eliminate air bubbles.

Fig. 32-14. Here is how the finished solar panel looks. Again, check the output voltage under sunlight. Connect the diode and put the solar panel to use in charging up those 12 volt batteries or operating projects.

sunlight. The solar panel is now ready to charge up any 12 V storage battery. Connect the enclosed diode to prevent the battery current from discharging through the solar panel. The positive terminal of the solar panel will connect to the positive battery terminal. Besides charging up auto, camper, or recreational batteries, you can power most 12 volt devices that have 750 mA or less current consumption.

PROJECT 33
CAMPER 12-VOLT SOLAR PANEL

A 36 solar cell panel may be used to charge those 12 volt batteries found in a recreational vehicle or camper (Fig. 33-1). You may even charge the battery in your favorite boat. In fact, this 1 amp 14 V solar panel will fully charge a 12 volt battery in one sunny day. Select lower-priced 3 or 4 inch cells for this project. Connect them in series, and you have a worthwhile solar panel with many uses. See Table 33-1.

Selecting the Cells

The solar cells for this project should produce from 1 to 2 amps. Since 36 large cells are needed, choose wisely. The three inch cell may be rated from 1 to 1.2 amps at .45 volts. A four inch cell may have a rating of 1.2 to 2 amps. Since the cells are rather expensive,

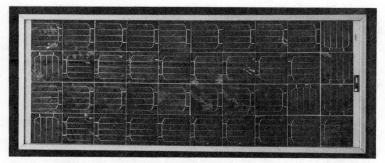

Fig. 33-1. Here is a commercial solar panel with forty series connected cells. It has a nominal output voltage of 17 volts with a 2.3 amp capacity (courtesy Solarex Corporation).

you may select manufacturer cutoffs, overruns or seconds. Check over the various cells and prices since they may vary from $8 to $15 per cell. Some firms may give added discounts when more than a dozen solar cells are purchased. So, look around. Shop for the correct cell with the required amperage and price.

Remember, 36 three inch cells will require a mounting area of 21 × 21 inches, while the four inch cells may require 12 × 44 inches for mounting. The three inch square cell may be mounted in a slightly smaller area. These larger cells may have single or double grid lines (Fig. 33-2). When double grid lines are found, both lines are soldered to the next cell.

Mounting the Cells

Layout the chosen cells in either a row of 4 or 6 cells. A 28 × 12 inch area is needed just to mount 36 three-inch cells. Of course, a larger mounting area is needed for four inch cells. These cells are mounted in a row of four cells with a total of nine rows. A square frame area may be obtained by placing six cells in a row with a total of six rows of cells. Whatever method you use, make a complete cell layout chart (Fig. 33-3).

Table 33-1. Parts List for Project 33.

Solar panel kit #128-B, purchased from Solar Electric Engineering, Inc., 438 West Cypress St., Glendale, CA 91204.
    All data and material courtesy of Solar Electric Engineering, Inc.
Cost of project: around $300.00.
Time of Project: 12 to 24 hours.

Fig. 33-2. You may find 1 or 2 grid lines on the large 3 or 4 inch solar cells. Both grid lines should be connected to the next cell for series hookup.

The cells must be mounted on a solid area so they will not crack or break. A lightweight panel may be constructed from sheet aluminum or galvanized sheet metal. The sides of the panel trough are strengthened with overlap metal. A piece of masonite material provides added mounting strength to the metal container. If possible, choose the sheet of aluminum over galvanized metal since it will not rust or need painting. The ½-inch metal trough may be constructed at the local sheet metal shop or you can do it yourself.

With four 3-inch solar cells mounted in 6 rows, the inside mounting area of the trough must be no smaller than 13 × 28 inches (Fig. 33-4). Since the trough is ½ inch deep with a ½ inch overlap, you need a piece of metal 15 × 30 inches. The overlap edge should

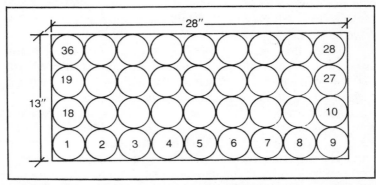

Fig. 33-3. Make out a layout chart for mounting the 36 solar cells. Draw the mounting area right on the masonite panel or on a separate piece of cardboard.

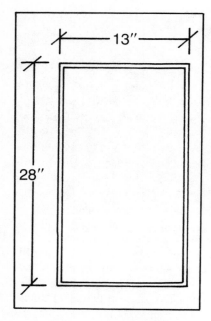

Fig. 33-4. The four cells in a row with a total of nine rows should not have a smaller mounting area than 13 × 28 inches. Instead of a rectangular solar panel you may want to end up with a square panel with six cells in a row.

be on the outside of the container. All corners are overlapped and bonded together with pop-rivets. The ¼ inch piece of masonite is cut to fit down inside the trough area. The masonite will help insulate the cells from the back metal area and provides added strength to the solar panel.

Cut the piece of masonite to the correct inside dimensions of the trough area. Sand down the edges so the piece will fit. Glue the piece of masonite to the bottom pan area. Fill the circle around the outside (between masonite and metal) with clear silicone rubber cement. This will prevent liquid plastic or rubber from running down inside the pan area.

Prepare the voltage output terminal strip before any cells are mounted on the masonite board. An insulated bolt and nut combination with dog-ears may be used for this purpose. But, a speaker pushbutton two-terminal strip is ideal for voltage take-off. These two speaker type connections come mounted on a piece of bakelite with two mounting holes (Radio Shack 274-315). Mount the terminal strip on the front side, one inch from the edge. Bend the metal terminal tabs out flat and mount the strip with ¼ inch metal spacers. Drill two holes through the masonite and metal backing. Bolt the strip in place (Fig. 33-5). The two metal tabs are inserted on top of the masonite panel. Now the voltage cable can be connected by pushing down on each button and sticking the wires through for easy hookup.

Connecting the Cells

Before mounting the cells haphazardly, layout the cell mounting on the masonite panel or make a cell layout chart from a piece of heavy cardboard. Cut the cardboard the same size as the masonite panel. Lay four cells evenly spaced at the top and bottom of the panel area. Draw a line on each side of the cells. Now, evenly space nine cells lengthwise on each side of the masonite panel. Draw lines crosswise. The cells will be mounted inside these square areas. If you like, place the cell in each square and draw a circle around it for the entire 36 cells.

All cells must be connected in series (Fig. 33-6). Start at the terminal area and number each cell. The cells are wired in a crosscross fashion until all 36 cells are wired in series. You will have four rows of cells with 9 cells in each row. Cells number 1 and 36 are the end cells for voltage take-off. The wiring from these cells is connected directly to the terminal strip. Since the final wiring is taken from the bottom of cell number 36 this is the positive terminal. Make sure this wire goes to the red lug of the terminal strip. The black pushbutton terminal connects to cell number 1 and is the negative terminal.

The solar cells may be all connected together from the piece of cardboard guide and then transferred to the masonite panel or they

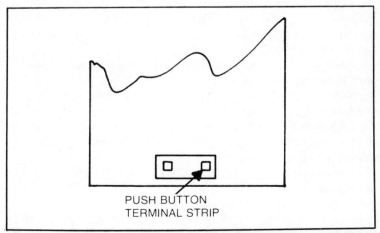

PUSH BUTTON
TERMINAL STRIP

Fig. 33-5. A speaker type pushbutton terminal strip is ideal to use as a voltage take-off terminal. Mount the terminal strip at one end of the solar panel.

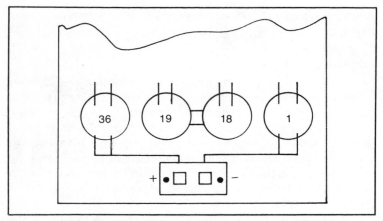

Fig. 33-6. Here all 36 solar cells are connected in series. Cells number 1 and 36 are connected to the terminal take-off strip. The black button is the negative and the red button is the positive terminal connection.

may be mounted separately. When only one grid line is used the cells may be mounted separately. Sometimes it's difficult to keep the grid lines straight when the cells are laid flat on the face side. Also, the cells have a tendency to float up when liquid plastic or solder is poured in the area. You must keep pushing down on the cells to help air bubbles escape from underneath. Sometimes these air bubbles are difficult to remove if the liquid plastic sets up too soon. To prevent the cells from floating, a dab of rubber cement may be applied to the bottom side of each cell, or the cells may be mounted separately with a dab of cement under them. With this method you can mount the cells directly on the masonite panel and keep every cell uniform.

When wiring all cells together connect a 1 inch piece of hookup wire to each grid line of the front cell side. Solder two connecting wires to the double grid cells. Use bare hookup wire for each connection. Of course, flat nickel wire is ideal, if handy (Fig. 33-7). Remove the insulation from regular hookup wire before soldering to the top side of the cell. Connect all cells in the very same manner except cell number 1 and 36. Tie a 10″ wire to the bottom of cell 36 and the top side of cell 1. A short 1 inch piece will solder to each connecting wire on double grid line cells. These long wires will solder to the output voltage terminal strip.

After connecting wires to the front of each cell, lay the cells face down on the guide card. Make sure cell number one has the 10 inch lead connected to the top of the cell. Cell number 36 will have a

Fig. 33-7. After all double grid cells are connected together they should look like those in this photo. Keep all connections as neat as possible.

10 inch wire connected to the bottom side. Lay each cell in line to connect to the next cell. The two connecting wires of cell 9 will go up and connect to cell number 10. Likewise, cell 18 will go upward and connect to cell number 19 until all rows are wired together. Place a layer of scotch tape on the back of each four cells in a row (Fig. 33-8). This will help keep the cells together when turning them over and transferring the cells to the masonite panel.

When the cells are mounted separately on the masonite panel, the bottom connecting wires must be soldered before the cells are

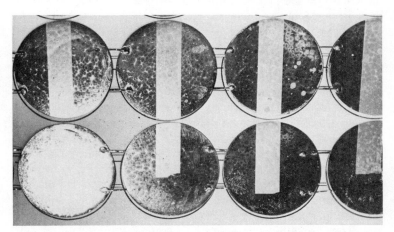

Fig. 33-8. Each row of cells is held together with scotch tape. This holds the cell array together while transferring it to the masonite panel.

mounted. This means the top grid lines must be marked from the bottom side of each cell, so the connecting wires are in line. This is easily done by marking each grid line on the bottom side with a pencil mark.

Now, solder the two connecting wires to the bottom side of each cell. Each cell may be moved and turned before the cement sets up. Sometimes when connecting the cells from the back side you cannot see if the grid lines are in line. This process takes a little longer, but when finished, each cell is in line and uniform in all directions.

Place a dab of cement on each cell and place it in the squares on the masonite panel. Keep all connecting grid lines in order. Lightly press down each cell. Pull the connecting wires up and over the top of the next cell. The connecting wires of each cell may be soldered to the next cell or leave the cells until the cement sets up. Then connect the two connecting wires to each cell. Try to make a neat flowing solder connection from the topside. This side will always be visible through the liquid plastic. Remember, too much heat may lift the soldered grid line. Use a low wattage soldering iron for solar cell connections. Snip off the wire ends if they are too long for a neat appearance.

Testing

Before liquid plastic or rubber is applied to the top of the cells, take a voltage test. Set the solar panel in the sunlight and measure the voltage at the take-off terminals. Without any load you should have a voltage measurement of about 16 volts. Usually, the voltage varies between 16 and 19 volts. When the output voltage is lower than 16 volts, suspect a cracked cell or poorly soldered connection.

Take a voltage measurement across two cells at a time until you notice a low reading. Compare this voltage with each set of cells. A very low reading will indicate a defective cell or poorly soldered connection. You may take a resistance measurement across each cell to determine which cell is defective. Of course, the cells should be out of the sun for any resistance check. Measure the resistance from the front of one cell to the same connection of the next cell (Fig. 33-9). Most large 3 or 4 inch cells have a resistance of less than 50 ohms. The ideal cell resistance connection is around 10 ohms. Check the suspected cell for cracks or a poorly soldered connection. Remove and replace any defective cell. You do not want any defective cell in this series hookup or it will prevent proper voltage and current output.

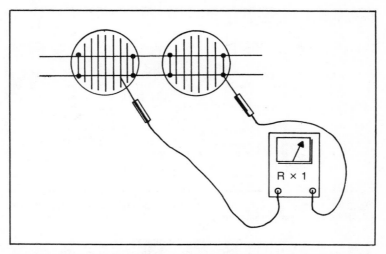

Fig. 33-9. After a defective cell or a poor connection is located with voltage tests, a resistance measurement may point out the defective component. Measure the resistance from the front of each cell to the connecting wire of the next cell. An ideal reading is around 10 ohms or less.

Finishing the Project

To prevent damage to the solar cells, several coats of plastic resin, a plastic cover, or poured liquid plastic may be placed over the solar panel. The cheapest method is to apply several coats of plastic resin over the cells. For additional protection place a piece of ribbed plastic over the entire cell area. For an ideal solar panel, apply liquid plastic over the entire panel area. The liquid plastic should be at least ¼ inch thick.

Mix the liquid plastic and hardener outside before pouring it over the panel surface. The plastic may be poured almost to the top of the metal edges. Make sure the panel is on a level surface. You may encounter some strong fumes from this mixture. Spread the liquid resin with a small brush. Make sure the liquid seeps under and around all cells. Use a brush to eliminate any air bubbles. Sometimes air bubbles may appear when the panel is setting up. Difficult air bubbles may be eliminated by puncturing with a tooth pick. The plastic may set up within 30 minutes in the summer months. When the temperature is lower than 70 degrees it may take a couple of days. You may place the solar panel near a heat register after the first day to advance the process.

A regular 2.5 amp silicon diode may be installed in one leg of the solar panel so the battery will not discharge through it. Now the

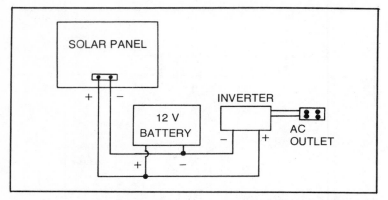

Fig. 33-10. A typical solar panel, inverter, and battery hookup may provide ac power for those small power tools.

Table 33-2. Commercial Solar Panels.

Arco Solar, Inc., 20554 Plummer St., Chatsworth, CA 91311
Applier Solar Energy Corp, 15251 E. Don Julian Road,
P.O. Box 1212, City of Industry, CA 91749
Motorola, Inc., 5005 East McDowell Road, Phoenix, AZ 85008
Silicon Sensors, Inc., Hwy 18 East, Dodgeville, WI 53533
Solar Amp, Inc., P.O. Box 27885, Denver, CO 80227
Solarex Corp, 1335 Piccard Drive, Rockville, MD 20850
Solar Energy Co., 810 18th St., Washington, DC 20006
Solar Systems Inc., 8100 Central Park, Skokie, IL 60076

solar panel may be connected to a 12 volt battery for charging purposes. A 12 volt inverter may be connected to the battery to operate small power tools or a TV when ac power is not available (Fig. 33-10). These 12 to 115 volt ac inverters come in 75 to 300 watt units. Radio Shack has one at 300 watts (22-130). Terdao Corporation has several models from 75 watts on upward. Heathkit has a power inverter kit for less than $55.00.

You may want to purchase commercial solar panels instead of building them. Of course, you are going to miss out on a lot of fun. Building things with your own hands can be a very rewarding experience. A list of firms selling solar panels is shown in Table 33-2.

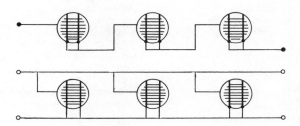

Suppliers

Mail order Electronic firms supplying low-priced, broken, or crescent solar cells.

Edmund Scientific
101 E. Gloucester Pike
Barrington, NJ 08007

John Meshna, Jr. Inc.
P.O. Box 62
East Lynn, MA 01904

Poly-Paks, Inc.
P.O. Box 942
S. Lynnfield, MA 01940

Solar Amp
P.O. Box 27885
Denver, CO 90227

Johnson & Smith Co.
35075 Automation
Mt. Clemens, MI 48043

Where to purchase larger solar cells (listed alphabetically).

Edmund Scientific
101 E. Gloucester Pike
Barrington, NJ 08007

ETCO Electronics
North Country Shopping Center
RT 9 North
Plattsburgh, NY 12901

GC Electronics
Rockford, Illinois 61101

GMI Electronics
715 Armour Rd.
North Kansas City, MO 64116

H & R, Inc.
401 E. Erie Ave.
Philadelphia, PA 19134

John J. Meshna, Jr., Inc.
P.O. Box 62
East Lynn, MA 01904

Poly-Paks, Inc.
P.O. Box 942
S. Lynnfield, MA 01940

Radio-Shack (at local stores)

Silicon Sensors, Inc.
Hiway 18 East
Dodgeville, WI 53533

Solar Amp, Inc.
P.O. Box 27885
Denver, CO 80227

Solar Electric Engineering, Inc.
438 West Cypress St.
Glendale, CA 91204

Electronic firms listed under parts list.

Edmund Scientific
Edmund Scientific Co.
101 E. Gloucester Pike
Barrington, NJ 08007

ETCO
ETCO Electronics
North Country Shopping Center
Rt. 9 North
Plattsburgh, NY 12901

GC
GC Electronics
Rockford, IL 61101

GMI
GMI Electronics
715 Armour Rd.
North Kansas City, MO 64110

H & R
H & R, Inc.
401 E. Erie Ave.
Philadelphia, PA 19134

Meshna, Inc.
John J. Meshna, Jr., Inc.
P.O. Box 62
East Lynn, MA 01904

Poly-Paks
Poly-Paks, Inc.
P.O. Box 942
S. Lynnfield, MA 01940

Radio-Shack
(at local stores)

Silicon Sensors
Silicon Sensors, Inc.
Highway 18 East
Dodgeville, WI 53533

Solar Amp
Solar Amp, Inc.
P.O. Box 27885
Denver, CO 80227

Solar Electric
Solar Electric Engineering, Inc.
438 West Cypress St.
Glendale, CA 91204

Magazines with advertisements for solar cell components.

Amateur Radio (CQ)
76 North Broadway
Hicksville, NY 11901

Elementary Electronics
380 Lexington Ave.
New York, NY 10017

Mechanics, Illustrated
1515 Broadway
New York, NY 10036

Popular Electronics
One Park Ave.
New York, NY 10016

Popular Mechanics
224 West 57th St.
New York, NY 10019

Popular Science Monthly
380 Madison Ave.
New York, NY 10017

Radio-Electronics
200 Park Ave. So.
New York, NY 10003

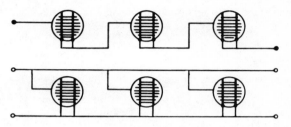

Index